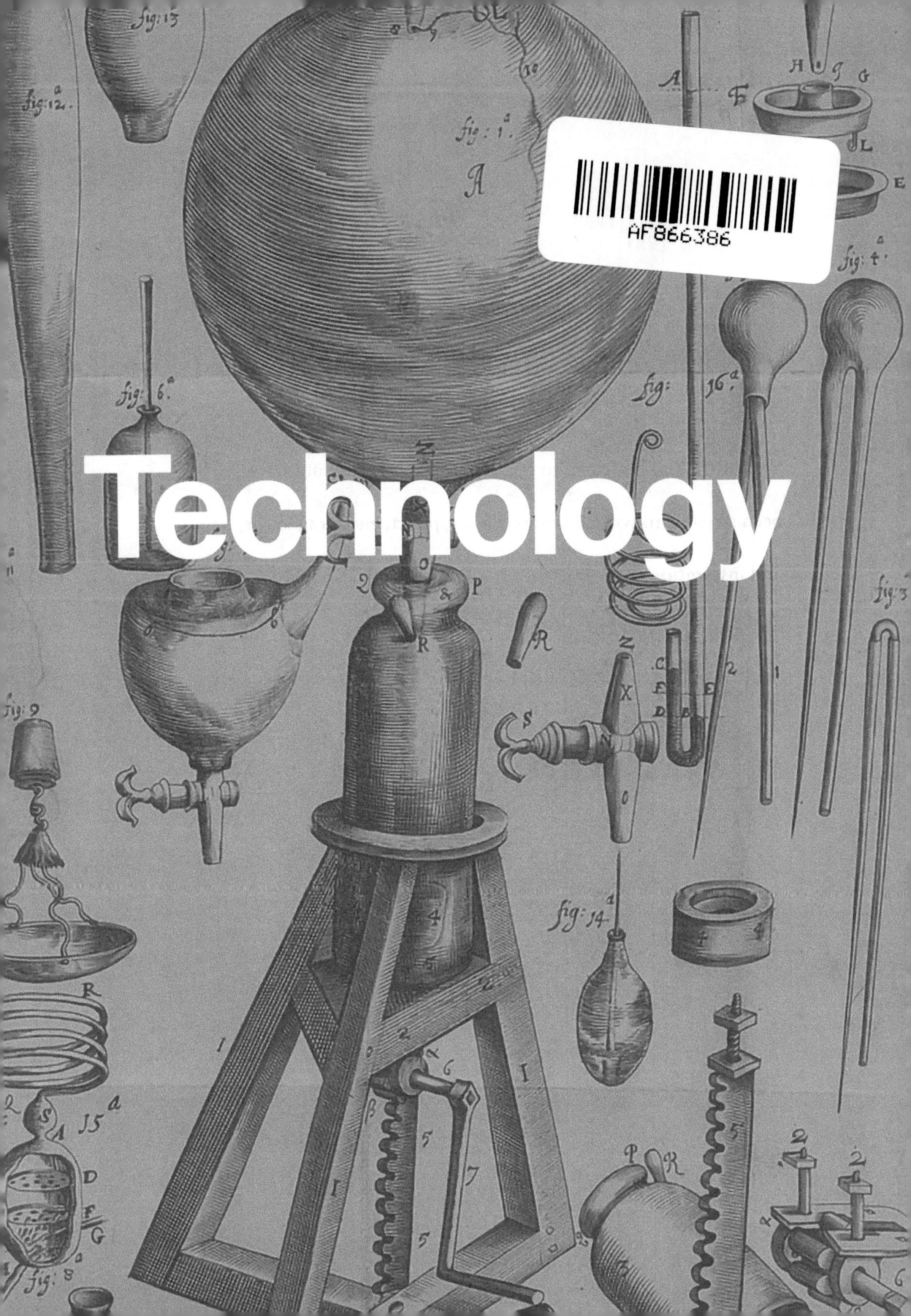

Technology

THIS BOOK is dedicated to my wife Betty. It would not have been possible to write it without her support, encouragement, and inspiration. I am grateful for all that we have shared for our over sixty years together.

Technology: Its Nature and Significance

William B. Jones

broadview press

BROADVIEW PRESS
Peterborough, Ontario, Canada

Founded in 1985, Broadview Press is a fully independent academic publishing house owned by approximately twenty-five shareholders—almost all of whom are either Broadview employees or Broadview authors. Broadview is supported by a collaboration with Trent University, a liberal arts university located in Peterborough, Ontario—the city where Broadview was founded and continues to operate. Broadview is committed to environmentally responsible publishing and fair business practices.

Library and Archives Canada Cataloguing in Publication

Title: Technology : its nature and significance / William B. Jones.
Names: Jones, William B. (College teacher), author.
Description: Includes bibliographical references and index.
Identifiers: Canadiana (print) 20260219444 | Canadiana (ebook) 20260219452 | ISBN 9781554810833 (softcover) | ISBN 9781770486171 (PDF) | ISBN 9781460405765 (EPUB)
Subjects: LCSH: Technology—Social aspects. | LCSH: Technology and civilization. | LCSH: Technology—History. | LCSH: Technological innovations—Social aspects.
Classification: LCC T14.5 .J66 2026 | DDC 303.48/3—dc23

Broadview Press handles its own distribution in Canada and the United States:
PO Box 1243, Peterborough, Ontario K9J 7H5, Canada
555 Riverwalk Parkway, Tonawanda, NY 14150, USA
Tel: (705) 482-5915
customerservice@broadviewpress.com

Broadview Press books are imported and distributed in the United Kingdom and European Union by:
Gazelle Book Services Ltd.
White Cross Mills, Hightown, Lancaster, Lancashire, LA1 4XS
sales@gazellebookservices.co.uk

European Union – Responsible Person (for official use only):
eucomply OÜ
Pärnu mnt 139b14
11317 Tallinn, Estonia
hello@eucompliancepartner.com
+33757690241

Broadview Press acknowledges the financial support of the Government of Canada for our publishing activities.

Canada

Edited by Archie Fields III and Michel Pharand
Book design and composition by George Kirkpatrick

Broadview Press® is the registered trademark of Broadview Press Inc.

PRINTED IN CANADA

1 2 3 4 5 6 7 8 9 10 26 27 28 29 30

Contents

Illustrations

Part One: Introducing Technology

Chapter I: Introduction

NOWADAYS ANYONE REFERRING TO "technology" is likely to mean computer and communication technology, or "high tech," as it is often called. Older devices and processes tend to be taken for granted, are often not even recognized as being technologies, and in many cases are all but invisible. Thus, there is a real danger that those grappling with the problems currently being generated by ongoing technological innovation will be unaware that the present situation is but the latest in a series of similar episodes, each of which has altered fundamentally not only the way our ancestors met their physical needs but also the way they thought about themselves, the world, and their place in it. And, of course, if one is not aware of these past episodes, one is in no position to benefit from whatever insights may be gained from studying them. In addition to preparing one to deal more adequately with the effects of current technological innovations, studying past examples that have contributed significant elements of human life as it is now lived helps one to better understand who one is, as directed by that ancient Delphic injunction, "Know thyself." The suggestion that understanding technology has a place in the pursuit of self-knowledge may seem farfetched, but any such doubts will be dispelled quickly by thoughtful consideration of the extent to which not only the physical objects with which one interacts daily but also one's thoughts and experiences are produced or formed by technology. A third reason for studying these past episodes is that technology, as many scholars have come to recognize, is not one thing but many things. Its bewildering variety does not lend itself to being subsumed under grand, sweeping theories but must be confronted in the nitty-gritty of practice. Only by this route can one gain an understanding of technology and an appreciation of the imaginative and inventive powers of the human species—as well as an understanding of the problems created. Hence the first part of this book, "Introducing Technology" (Chapters I and II), focuses on setting out the key issues to be discussed throughout the book as well as providing a definition of technology in which to anchor that discussion.

What besides "high tech" is to be included under the heading "technology"? That is the first question faced by anyone who sets out to learn more about technology and to reflect upon its significance. The term "technology" does not designate a category long recognized in human language, nor does it label a particular kind of object or process that can be pointed to or described in a simple or straightforward manner, such as a newly discovered plant species. Rather, this usage has resulted from a growing recognition that there is an important aspect of the human scene (in a sense comparable to politics, warfare, or religion) that merits attention and study. Long before technology was recognized as a distinct and significant category, the effects of technological innovation captured the attention of writers, thinkers, and social critics. Given the power of some technologies, it is inevitable that their implementation would be disruptive and produce problems, social and/or environmental ones. Writings from ancient times express concerns about some of these, but it was as the Industrial Revolution proceeded that voices became shrill and violence erupted in opposition to the introduction of new means of production. Even then, the targets of the social criticism—and of the violence—were specific types of situations, trends, or innovations.

But by the beginning of the twentieth century, there was a growing awareness that something fundamental and very powerful was at work. Social critics, seeking to identify the culprit, pointed in the direction of technology. But they put forward very different accounts and used different terms and categorizations. Confronted by such horrors as powerful new explosives used in warfare, chemical wastes, industrial slums, mustard gas, and nuclear weapons, some thinkers have argued that science and the kind of thinking that it involves is the genie that has been let out of the bottle. Accordingly, they have tended to draw a fundamental distinction between recent science-based technology and the traditional practical arts. On this point, their position is in accord with the once-popular view that technology, properly understood and efficiently practiced, is "applied science," i.e., scientific knowledge applied to practical problems. However, these thinkers advance it as an indictment rather than a recommendation.

Modern readers may find it surprising that science and technology have not always gone hand-in-hand. Science and technology are now such a powerful partnership that few people are likely to be aware that the two have not always advanced together or been as mutually supportive as they currently are. Indeed, most people probably do not make a clear distinction between technology and science. However, it was not until the nineteenth century that scientific knowledge, as such, became a major factor in technological innovation. Much earlier—during the seventeenth century—modern science burst upon the scene with successes made possible by its technological resources: Tycho Brahe's giant quadrants, Galileo's telescope, his bronze sphere rolling down a sloping grooved track, Torricelli's barometer, Leeuwenhoek's microscope, and other inventions. Prior to all of this, the two enterprises developed in relative independence of each other. Making scientific instruments is a vital part of the scientific enterprise, and much of this work is performed by scientists themselves. Thus, while modern science is intimately related to technology, that has not always been the case.

Others have traced the problems faced by modern society back to the Industrial Revolution or even earlier. These differing views reinforce the point that any serious discussion of technology and issues related to it must begin with a systematic consideration of what is to be included under the heading "technology," including its relation to science, the topic of Chapter II. Questions of this kind, i.e., regarding just which things are to be included in a class or category, are termed "conceptual" issues. Such matters are a central concern of philosophy.

With a definition of technology in hand, one can begin to inquire about its role in the human drama. To begin with, there is compelling scientific evidence (1) that neither human beings nor technology have always existed on this planet, (2) that over time both of them emerged from antecedents, and (3) that fully modern human beings and technology (as defined in Chapter II with an emphasis on innovation) appeared together some one hundred thousand years ago. The second part of this book, "Technology and Human Development" (Chapters III through VI), is broadly concerned with illustrating key episodes in the co-development of humanity and technology. And while the chapters proceed roughly chronologically, occasional leaps ahead are made to illustrate important concepts concerning the historical development of technology.

Making and using tools was essential to the biological success of the antecedents of modern human beings and to their taking up residence beyond the confines of Africa. According to archaeological evidence, there seems to have been a series of periods in which tools/weapons were more sophisticated and effective and their makers more capable in relevant ways than during a previous period. During each of these intervals the relevant human ancestor enjoyed advantages not available to its predecessors, advantages deriving from tools and techniques dependent upon their own increased capabilities. This process culminated in the appearance of both fully modern human beings (with their inventiveness, aesthetic expression, use of symbols, bodily ornamentation, complex social organization, long distance trade, and so on) and technology. The highlights of these developments are sketched in Chapter III.

Given the importance of technological innovation, the question of what factors affect it immediately rises. At the most fundamental level, the degree of technological development that people are likely to attain is heavily dependent upon having a more-or-less permanent place of residence. If one's possessions are not limited to what one can carry about, then one can have a much greater variety of tools and materials at one's disposal. Having a large variety of specialized tools enables one to carry out a great variety of operations and to utilize materials that one could not otherwise make use of. A special advantage of sedentism is that it allows materials that are available only at certain times of the year to be warehoused and used in combination with other materials to produce a single product. Most importantly, since most new technologies derive from old technologies (through new uses, new combinations, and the realization of possibilities that would not have existed in the absence of the old technologies), the greater the variety of tools and techniques people have, the greater the potential for further technological development. This fundamental feature of technological development may be compared to rolling a snowball: the bigger it is, the more snow it picks up with every rotation. Even in the absence of agriculture, at certain places there have been times when seasonal abundance of food sources has made it possible for people who develop ways of storing this bounty to maintain a permanent dwelling place. Examples are discussed in Chapter IV, which features a case study of the impressive technologies of the Native Americans of the Pacific Northwest.

Although a sedentary or semi-sedentary way of life is possible in the absence of agriculture, the converse is not true: the practice of agriculture imposes or reinforces (to one degree or another) a sedentary way of life upon those who engage in it. If one is to benefit from planting a crop, one must, at the very least, return to the site to harvest it and may need to be present during the interim, for example to drive off marauding animals or birds. Furthermore, if the harvested crop constitutes a significant fraction of one's yearly food supply, it will be far too massive to be carried about and will have to be stored—and probably guarded. Thus, those who avail themselves of whatever advantages agriculture has to offer must adopt a way of life that is, more or less, sedentary. Furthermore, there is a growing body of evidence suggesting that the practice of planting crops originated among people who were already following a sedentary or semi-sedentary way of life. It is people in just such circumstances who can benefit from, as well as initiate, incremental improvements in the quantity or quality of foodstuffs obtained from plants. In recent decades, we have learned a great deal about the development of agricultural practices capable of providing all or most of people's nutritional needs. The steps taken in the Ancient Middle East to achieve this end have been extensively studied, hence the topic makes an excellent case study.

Not surprisingly, many of the effects traditionally associated with the advent of agriculture are more directly related to sedentism than to farming. Among the developments more closely connected to the practice of agriculture *per se* is a dramatic increase in the human population. On average, women living in agricultural societies have a much larger number of children than is typical of foragers. Farming communities can produce food soft enough for children to eat at an earlier age than that at which the children of hunter-gatherers are able to consume foods available to them other than their mothers' milk. Thus, the women of agricultural societies do not have to nurse their children nearly as long or as intensely as do women in foraging societies, who may nurse theirs for as long as four years with the result that their children are born further apart. Furthermore, in agricultural societies children (even at an early age) can perform quite a variety of useful tasks and consequently are an economic asset. Children continued to play a significant role in the economic activities of their communities until the advent of the Industrial Revolution, when the locus of much productive work (especially that performed by men) began to shift away from people's homes (or adjacent workshops).

The inevitable results of agricultural peoples' having large families include a burgeoning population and the need for more land. Farming populations grow and spread, either absorbing or displacing groups of hunter-gatherers who are encountered. This process explains one of the most striking features of the human cultural scene: the existence of closely related languages, "language families," spoken by people spread across vast areas. The suggestion is that these languages are similar in vocabulary and grammatical structure because they are "daughter" languages, languages that diverged from an ancestral tongue as the agricultural way of life spread more and more broadly. An excellent example is provided by the Bantu languages, which are spoken by the inhabitants of a vast swath of central and southern Africa. If the proposed explanation is correct, these languages are descendants of a language once spoken by a farming people living in what is now southeastern Nigeria and western Cameroon. Recent studies have yielded insights into the development not only of these languages but also of African agriculture. Thus the "Bantu expansion," as it has been called, is an excellent choice for a case study for Chapter V.

The importance of the effect that agriculture has had upon language must not be underestimated. It has been said that "All the soul of a people is in its language," and students of foreign languages are regularly reminded that one cannot learn a language only but also its culture. Accordingly, anyone who truly wishes to know "Who am I?" and wants to find one's roots must seek out the origins of one's language and identify the factors that contributed significantly to its being what it is. The developments sketched above make it clear that the agricultural revolution is one of these factors. The agricultural revolution and its implications for the languages spoken by many millions of people (and all that implies) illustrate an important general point: technology is not a mere means, a tool that serves the pre-defined ends of a completely pre-formed human nature, but an expression of human inventive powers which plays a central role in the ongoing creation of the very fabric of human life.

One of the most formidable barriers to understanding technologies and their development is the presumption, on the part of those familiar with a technology and its uses, that it was for just such purposes that the technology was devised. And while this assumption may be warranted in some cases (the cotton gin, the telephone, the incandescent bulb, among others), it leads to a fundamentally flawed overall view of how complex technologies like agriculture develop. Typically, they emerge over an extended period of time in a series of steps, each one realizing possibilities not likely to have been anticipated prior to the previous step. Viewed retrospectively, these developments may seem to have been successive advances toward an obvious goal, but they actually may have occurred quite independently of each other. The practice of agriculture was not and could not have been initiated in order to secure the benefits now believed to derive from it. As obvious as these advantages now seem, there is simply no way that the earliest farmers could have anticipated either the benefits that advances in agriculture would bring or the indirect benefits that the agricultural way of life would make possible. The emergence and development of agriculture is a much more complex matter than one might expect, as will be seen in Chapter V.

A second—and even more formidable—barrier to truly appreciating the significance of past technological innovations is the near-impossibility of imagining what it would be like to be without the technologies introduced, not just to live without the devices and techniques but to think like someone who is unacquainted with them and their uses. There could hardly be a more quintessential example of this inability than the invention of writing and numerals. Since it is nearly impossible for literate persons to imagine what it would be like not to have the use of writing and simple mathematics, it is extremely difficult for them to appreciate the achievement that writing and number systems represent. Only thanks to a combination of fortuitous circumstances and the life-long dedication of quite a variety of scholars and scientists was this obstacle surmounted. The magnitude of the challenge they faced is powerfully illustrated by the absence, prior to the 1930s, of any real understanding of oral cultures and non-literate patterns of thought. Up to that point, scientists and other thinkers (themselves literate!) had just assumed that only differences in factual information, belief systems, ways of classifying things, and the like distinguish literate people from those who do not have the use of writing.

As long as this view prevailed, it was impossible for literate persons to recognize the profound changes that came about following the invention of writing. There had to be, in effect, a rediscovery of non-literate thought. Once this had been achieved, it became possible to recognize what a profound transformation the "literacy revolution" really was: it involved a restructuring of thought processes, changes in the kind of thinking done, and alterations in how experience is organized. Most importantly, for science, mathematics, and philosophy, the literacy revolution involved the development of abstract thinking, the use of categories corresponding to the very general properties of things, geometrical shapes for example, in contrast to practical and concrete types such as wheels, doors, and mugs. Formal reasoning, such as one finds in geometrical proofs, was also involved. Unfortunately, only linguists and related specialists seem to be aware of these important developments.

Just as investigators have made great strides, beginning in the 1930s, toward understanding the significance of the invention of writing, so have researchers made remarkable progress, beginning in the 1970s, not only toward the discovery of the origins of elementary mathematics, as we know it, but also toward clarifying its very nature. Contrary to all reasonable expectations, investigators have been able to identify, based on concrete physical evidence, a series of steps taken by the Ancient Sumerians as they sought to meet the record-keeping needs of their increasingly complex society. The cumulative result of these steps, spanning more than a thousand years, was nothing less than the invention not only of writing but also of number systems as we know them. In this long process, the key elements of number systems—features now taken for granted—made their appearance one after another in relative independence of each other. That they did so makes it clear that some aspects of number systems which might appear to be simple and elemental are neither. It would be difficult to overemphasize the economic, political, social, scientific, and aesthetic significance of the development of number systems and the ability to record both quantitative and qualitative information. Chapter VI is devoted to these topics using the development of Sumerian record keeping as the case study.

The third part of this book, "Technology and the Modern World" (Chapters VII through IX), explores important relationships between technology and other central spheres of the modern human world: economics, society and politics, and the environment. Since these chapters each characterize a key relationship between technology and some other domain, we use different historical eras and episodes to illustrate these relationships. The reader must therefore allow for departures from strict chronological ordering for the sake of making connections between important events.

Given the definition of technology elaborated in Chapter II, one can see how technology is closely related to economic activity, the production of goods, and the rendering of services. This connection is exemplified by, among others, the textiles and giant dugout canoes of the Native Americans of the Pacific Northwest, Bantu agriculture, and Sumerian record keeping. In Europe, economic activity received an especially important boost from technological innovation during the Late Middle Ages. During this period, technological advances accelerated to the extent that historians speak of a "Medieval Industrial Revolution." Some regard the Industrial Revolution of the eighteenth and nineteenth centuries as the continuation of processes set in motion during this earlier period. Yet, as strange as it now seems, the economic importance of technology (as such) was generally recognized only after the Second World War.

Of course, in the modern world, technology is intimately connected to science, as discussed in Chapter II, and so the impact of economic factors on science must also be considered. Not only is scientific knowledge often of practical/economic value, but the development of instruments for scientific purposes repeatedly has paved the way to practical economic benefits. Furthermore, since costs in time and money inevitably influence which lines of research are pursued, the practical/economic promise of a technology being developed for scientific purposes may attract support for that research program irrespective of its scientific goals. Evidently, there are important questions needing to be addressed regarding technology, economics, and science and how they relate to each other. Chapter VII is devoted to these topics.

Questions about how technology and science relate not only to economics but also to politics and other societal elements have been the subject of much study and debate, generating an immense body of literature. Although some concern for the effects of one or another technological practice was expressed in ancient times, it was only with the Industrial Revolution that the impact of technological innovations became a central theme of social criticism. Also notable during this period were organized attempts, by those negatively affected by new technologies, to prevent their implementation. Protests sometimes escalated into mob violence and large-scale riots. One particularly prominent grievance was the introduction of power-driven machines that produced goods faster, more cheaply, and in greater quantities than traditional methods and could be tended by low-paid operators with minimal training. Such developments threatened the value of the special skills of traditional craftspeople, their way of life, and their very survival. Situations of this kind bring into sharp focus the ambiguous character of technological innovation: new devices and techniques may make it possible to produce articles with less labor, thus increasing their availability, but they often render worthless the skills and experience of large numbers of people and may enrich one group or render it more powerful at the expense of others. Technological innovations can also effect major changes in family life. As the Industrial Revolution proceeded, the role of fathers became increasingly that of a wage earner, while mothers tended to remain at home with, if anything, even more responsibilities. Chapter VIII includes a discussion of this and related topics.

Some have argued that the culprit responsible for the plight of the craftspeople displaced during the Industrial Revolution (and in similar cases) is not technology but the socio-political system (e.g., industrial capitalism) in which it was implemented. According to this view, called "instrumentalism," technologies are (in simplest terms) just tools, ways of getting things done, and any good or ill that results from their use derives from how they are used or the ends they are deployed to promote. The thesis that technology is in this sense "neutral" has been challenged by, among others, social scientists and historians who subscribe to what is called "social contextualism" or "social determinism." These thinkers point out that a technology can hardly be said to be "neutral" if it reinforces markedly the economic and political power of those who control it and introduces changes that are unwanted by a majority of the population. Social contextualists argue that the make-up of technologies derives not only from technical considerations of functionality but also from a variety of political, social, and cultural factors, including value systems. These thinkers are also critical of "technological determinism," according to which technology, rather than being guided by human goals and decisions, is an autonomous entity that follows its own course of development determined by its own internal logic. In this view, technology is like a large beast which people have managed to mount only to discover that they cannot control it.

Technological determinism has a certain initial plausibility, and there are numerous articulate proponents of one form or another of it. But its critics have called attention to many cases in which powerful technologies have failed to win acceptance, have been proscribed or tightly restricted, or simply have not been utilized. Noteworthy examples include the banning of the insecticide DDT in the United States and most other industrialized countries, combatants' refraining from using poison gas in the Second World War, and the virtual elimination of firearms from Japan for a 250-year period beginning in the seventeenth century.

This last example is particularly interesting. Firearms were introduced into Japan by the Portuguese in the 1540s and in short order were being manufactured there by the thousands before their numbers began to be reduced. In contrast, the introduction of gunpowder into Europe was followed in quick succession by radical changes not only in warfare but in many other areas as well. In China, on the other hand, refinements and applications of the "fire drug" (as gunpowder was known in the land where it was invented) proceeded in a step-by-step process extending over several centuries. Contrary to Western myths, these included military uses; however, no revolutionary changes followed. Clearly, an examination of the historical and cultural contexts of these three radically different outcomes (rejection of a technology, acceptance followed by revolutionary changes, and acceptance without revolutionary changes) enables a better understanding of how technology relates to other aspects of human life. Accordingly, it makes an excellent choice for a case study for Chapter VIII.

Although awareness of the human impact upon the natural environment dates back to at least to the Ancient Greeks (that soil erosion, resulting from land being cleared to make way for growing crops, had reduced Plato's native Attica to a "skeleton" of its former self is expressed in his *Critias* 111a–112a), and although organized efforts to preserve wilderness areas were getting under way in the U.S. in the latter part of the nineteenth century, the level of public concern for the welfare of the environment was not yet high enough to support substantial action to prevent its impending, irreversible impoverishment. Thus, for example, as the final destruction of the once-immense buffalo herds (as well as the possible extinction of the species) loomed in the 1870s, efforts to enact federal legislation to protect the American bison were stymied by those who had other priorities. The species was saved from extinction by the actions of only a few individuals.

Four technologies (the domesticated horse, the breech-loading rifle, the steam locomotive, and refrigeration) played crucial roles in the (near-) eradication of the buffalo in a complex set of developments involving demographic, economic, political, military, and socio-psychological factors. Because it is such a good example of how the effects of technological innovations are conditioned by such factors, and because the near-extinction of such an iconic species raised public awareness of the gravity of the threat of anthropogenic (human-caused) extinctions, the fate of the buffalo is ideally suited for a case study for Chapter IX. The second section of the chapter features four examples of the over-exploitation or misuse of a natural resource that were followed by the adoption of corrective measures: the extinction of the passenger pigeon, the devastation of the marine fisheries of the world in the second half of the twentieth century, the "dust bowl" of the 1930s, and the "killer smog" of October 1948 in Donora, Pennsylvania.

Just as the herds of buffalo that once blackened the North American plains were ruthlessly exploited to the point that extinction loomed, so were the flocks of passenger pigeons that once darkened the skies. Although the chain of events leading to this result is more complex than that which culminated in the destruction of the buffalo herds, two inventions did play key roles in the commercialization of wild pigeon as a food product and the decimation of the flocks: the breech-loading, double-barreled shotgun and the steam locomotive (which made it possible to move the meat rapidly over long distances to mass markets). In this case, unfortunately, public awareness and concern for the birds' rapidly declining numbers and actions taken by individuals and groups were not sufficient to prevent the extinction of the species. The attention generated, however, did lead to the adoption of laws, regulations, and practices that saved other avian species from a similar fate.

Developments in the marine fishing industry in the second half of the twentieth century provide an even more disturbing example of what can happen when increasingly powerful technologies are deployed in the pursuit of wild game and fish. Things came to a head in the third quarter of the twentieth century when an entirely new kind of fishing vessel, huge factory trawlers, began to appear wherever fish were abundant. These giant trawlers could scoop up enormous quantities of fish and both process and freeze them on board. Since there was no urgency to get their catch to port, these ships could operate anywhere in the world as long as necessary to obtain a full load. These fishing operations were so efficient that few fish were left for local fishermen or for reproductive purposes. In response, countries with coastlines adjacent to the good fisheries assumed jurisdiction over "exclusive economic zones" extending out 200 miles from their shores. Factory trawlers were excluded from or severely restricted within these zones. Since the good fisheries were located in the shallow waters of the continental shelves, the result was that these giant fishing vessels were no longer economically viable. Ironically, even if no restrictions had been imposed upon them, the factory ships would soon have sealed their own fate by destroying these fisheries.

Unfortunately, the retirement of the giant trawlers did not reverse or even halt the decline of marine fisheries. This point is powerfully illustrated by the fate of the Grand Banks and the Georges Bank, two of the richest fisheries in the world which lie off the southeast coast of Canada and New England. Technological innovations together with a growing number of fishing operations at first had resulted in an enormous increase in the tonnage of fish brought in. But this total began to decline around 1950, even before the arrival of the factory ships, a trend that continued after their departure. By the early 1990s, the decline was so pronounced that the US government declared a complete moratorium on fishing in part of the Georges Bank in order to give the population of desirable species a chance to rebound. For the same reason, Canada shut down the Grand Banks fishery even though 40,000 jobs were at stake. There could hardly be a better example of a technology being so powerful that its unrestrained use renders it completely useless.

The Donora incident is just one example of the disasters that focused attention, in the decades following the Second World War, upon the necessity of bringing air pollution under control and doing something about the abysmal state of the rivers, lakes, and streams. At mid-century, people continued, as they had for thousands of years, to treat rivers as waste-disposal facilities. But at this point the horrendous effects of such practices had reached unprecedented levels. This situation came about as the result of population increases, urbanization, industrialization, and an entirely new range of toxic wastes from the petroleum and chemical industries and from other production operations ushered in by the formation of the powerful partnership of science and technology. As environmental degradation assumed a place in the public consciousness, heated debates began to take place concerning the "roots" of the "environmental crisis." Initially attention was focused on the havoc wreaked upon the natural order by the mining, manufacturing, and waste-disposal practices characteristic of the Industrial Revolution and the chemical and electrical industries that sprang up in the late nineteenth century. Recent concerns include ozone layer depletion, acid rain, nuclear accidents, massive oil spills, and climate change.

In the context of such horrors, it can hardly be surprising that more traditional ways of living would have seemed relatively benign. But as the matter has been studied more systematically and more thoroughly, it has become increasingly clear that the human impact on the natural environment did not begin with the Industrial Revolution, the advent of metalworking, or even the development of agriculture. Increasing evidence shows that human beings were responsible for the extinction of species and radical changes in ecological systems at a much earlier date than had been believed previously. Indeed, we now know that states of affairs once thought to be examples of human beings living in harmony with nature are actually cases of people living in environments radically altered by human activity.

A fascinating and much debated topic is whether and/or to what extent predatory human beings were responsible for the extinction of large animal species in prehistoric times. This controversy was ignited by an article, "Pleistocene Overkill" (1967), in which such a hypothesis was advanced as an explanation for the puzzling disappearance in the Late Pleistocene of some two-thirds of the large-animal species of North America. The overall conclusion is that the impact of human beings upon the natural environment is roughly in proportion to their numbers and to the power of their technologies. Chapter IX enumerates the major technological developments and the negative impact upon the environment of the uses people have made of them.

The impact of powerful technologies upon the environment is one of three areas (the others being innovations in healthcare and computer and information technology) that have generated more interest, provoked more philosophical discussion, and occasioned more reflective input into decision-making than just about any other subject area related to technology. Concern about the deteriorating state of the environment and questions about the causes thereof and about human responsibilities regarding the environment have given rise to a new subdiscipline in philosophy: environmental ethics or, more broadly, environmental philosophy. The subject has proven to be particularly challenging because it is far from obvious how the welfare of animals, plants, species, ecosystems, etc. is to be fitted into traditional ethical systems, which (with few exceptions) have taken into account only the welfare or rights of human beings. As a result, philosophers addressing these issues have had to make a fresh start, yet again inquiring about the nature and extent of ethical obligation and about the place of human beings in the scheme of things, especially in relation to other living things with which they share the planet. Chapter IX includes a discussion of these topics.

Similarly, recent advances in medicine and related fields have forced people to make decisions not faced previously and to confront questions for which traditional views do not appear to provide ready answers. Furthermore, dealing adequately with the topics and questions involved often requires knowledge of biology, biochemistry, medicine, and other fields, knowledge that is quite specialized and advanced in nature as compared to that involved in discussions of most of the technologies treated elsewhere in this book. Because of this and also the need for the philosophical skills requisite to address questions of this kind, those concerned with these issues and willing to do what was necessary to address them set out on what proved to be a trailblazing undertaking. In doing so, they launched an applied ethics sub-discipline, biomedical ethics, which led to the establishment of applied ethics courses in college and university curricula beginning in the 1960s. Biomedical ethics (in one form or another) is now a well-established field in medical, nursing, and other health-related disciplines. Because of its specialized and advanced nature, however, biomedical ethics is not covered in this book. Rather, the interested reader is encouraged to consult one of the excellent texts on the subject, such as *Bioethics in Context: Moral, Legal, and Social Perspectives* by Gary E. Jones and Joseph DeMarco (Broadview Press, 2016).

The significance of computer and information technology derives not only from its great power but also from its having left so very few areas of human activity unaffected. Instead of enumerating these effects, which would be a huge task, the final, fourth part of the book, "Transition to the Digital Age" (Chapter X and Postscript), focuses upon one main element of the human scene: music. Chapter X traces the major changes that technological innovations have occasioned in how music is produced as well as heard and enjoyed. To provide a perspective on recent computer-related developments, the chapter begins with the earliest known musical instruments and techniques (in the West) and includes a discussion of early (mechanical) methods of recording and reproducing sound. The maturing of electronics and (subsequently) of computer systems and digitalization has made possible new modes of musical expression and even new genera as well as new modes by which listeners can access musical performances. The chapter also includes explanations of some of the scientific discoveries and inventions that made these developments possible, revisiting themes discussed throughout the book to better understand the history of music technology. A brief postscript concludes with a few thoughts on the future of technology in the digital age.

Chapter II: Defining "Technology"

IN SOME CASES, ONE can infer what is included in a given category simply from the term used to identify it. For example, "mineralogy" is the study of minerals. Similarly, "herpetology" is the study of reptiles and amphibians, although one could not infer that from the term "herpetology" alone unless one knew that *herpeton* is Greek for "reptile." Following this same pattern of thought, one might reasonably conclude that "technology" means the study of technical matters of one sort or another. And one would not be completely off the mark in doing so, as the expression derives from *techné*, Greek for craft, art, or skill, and it originally meant a study of such topics—albeit not the arts or skills themselves. However, in recognition of the importance of the powerful new devices and techniques then being developed (and their effects), and in response to the need for a convenient way to refer to them, the usage of "technology" broadened during the second half of the nineteenth century to include the practical arts themselves (collectively), as well as writings or treatments addressing such matters. This broadening appears to have taken place first in Germany, where industrialization was proceeding at a dizzying pace, spreading to North America only after the First World War. Further broadening enabled the use of "technology" to refer to specific technologies, for example, "petroleum refining technology."

During this period, philosophers, social critics, and others, disturbed by conditions then prevailing, first began to reach a wide audience with treatments in which technology, as a concept, figured prominently. However, since they had very different views regarding both the sources of and the remedies for the ills which they perceived, they conceptualized the situation differently and, consequently, included different things under the heading "technology." Thus, as pointed out in the Introduction, for a meaningful discussion of issues relating to technology to take place, one must come to an understanding of what "technology" is assumed to mean.

A. The Nature of the Question

First, it should be clear that the question here faced cannot be settled by consulting a dictionary. Dictionary definitions are reports of usages and often include several that may be incompatible or sufficiently different to cause problems if assumed to be equivalent. This point is aptly illustrated by "science." Using it as an example also provides an opportunity to initiate discussion and reflection about one of the central questions that must be addressed in any serious treatment of technology and its nature: How does technology differ from and relate to science? Conceptual issues of this kind are, of course, the province of philosophy. But while philosophical reflection specifically focused on science ("philosophy of science") has been ongoing since the middle of the nineteenth century, only in the twentieth century did philosophers and allied thinkers begin to make technology and its ramifications a central focus of their reflections.

According to one definition, "science" is an ongoing tradition of inquiry pursuing knowledge and understanding of phenomena, as exemplified by physics, chemistry, biology, geology, and other fields. A second meaning of "science" is a methodology for carrying out just about any kind of task, including packing for a trip. ("She has packing for one of these trips down to a science.") Although there has been much discussion (and disagreement) concerning the exact nature of this "scientific method," it typically has been portrayed as involving meticulous attention to detail and rigorous logic. This approach to problem solving (scientific or otherwise) has been deemed responsible for the ills of modern society and for stifling human spontaneity, creativity, and imaginative powers. According to those advancing such criticisms, it is the scientific approach that accounts for the power and the destructiveness of recent technologies (understood as instances of "applied science") and distinguishes them from traditional arts and crafts.

However, although it was once widely believed that the key to the development of modern science was "the scientific method," extensive studies carried out during the last half century or so have found that actual scientific work bears little resemblance to what such a methodology would lead one to expect. While there are elements of scientific endeavors that do fit this pattern, also essential are imagination, inventiveness, and innovative thinking. Furthermore, during this same period investigators have concluded that most new technologies derive from existing devices and techniques, not from the application of scientific findings—and that this has remained true even after scientific knowledge became an important resource for technology late in the nineteenth century. Accordingly, it is neither accurate nor illuminating to think of technology as "applied science." It follows that the relation between the scientific approach (the second sense of "science" identified above) and the harms and benefits bestowed by technology are, at best, tenuous.

Even so, it must not be forgotten that, in the latter part of the nineteenth century, science (in the first sense identified above), especially chemistry and physics, became an important resource for technology, providing the foundation for the chemical industry and the electrical and (electronic) communication industries. Unhappily, these technologies have been responsible, not only for numerous benefits now taken for granted by so many people, but also for some of the worst horrors of the modern world: powerful new explosives and poison gases for use in warfare and unprecedented forms and levels of pollution. Subsequently, advances in nuclear physics opened the door for nuclear weapons, nuclear power, and nuclear medicine along with radioactive wastes. Given the importance and visibility of such technologies, it is hardly surprising that they came to be regarded as paradigms of modern technology and its relation to science. The question of how technology relates to science clearly deserves attention, which it will receive in Section E below.

The second point that needs to be kept in mind is that much more is at stake than merely defining a word. Rather, the question is (a) whether there is a category or group of things, events, or processes of a certain kind that merits special attention and (b) if so, what should its boundaries be? By the early decades of the twentieth century, an impressive number of authors, social critics, philosophers, and others were answering the first question in regard to technology with an emphatic "Yes!" But, as noted, they did not give the same answer to the follow-up question.

The third point that merits attention at this juncture is that the question of what is to be included in the category "technology" is not a factual or empirical matter, as is, for example, the question of whether the class of pit vipers includes rattlesnakes. (Pit vipers are fanged, venomous snakes with sense organs, "pits," that respond to infrared radiation. Rattlesnakes do meet these conditions.) The statement is empirical because it asserts that the members of one of two well-defined classes are among the members of the other and because it is possible to verify the accuracy of the statement by observing, conducting tests, consulting records, etc. In the context of common usage there would seem to be little room to doubt that well-established classifications like "pit vipers" and "rattlesnakes" group things together on the basis of shared characteristics.

But languages and the categorizations they embody are not static. They change when people confront new types of things or situations, when they recognize the need to refer to aspects of things and relationships not previously conceptualized, when there is a need for more clarity and precision, when new ways of classifying things gain favor, and when empirical findings become so embedded in human thought that they are incorporated into definitions. And the need for such changes will arise as people learn more, endeavor to organize knowledge, seek to classify and understand phenomena, etc. Unlike "rattlesnake," "technology" does not denote a well-established category. It is a term that came into use as writers, thinkers, and social critics recognized and drew attention to an aspect of the human scene that they believed deserved special attention. In doing so, they introduced not only a new use of an expression but also a new (or at least an enlarged and revised) way of categorizing things. As noted above, questions and issues related to how things are categorized are said to be "conceptual" (as opposed to factual) in nature, and these include the question of what is to be included under the heading "technology." That is the fourth and most important thing to understand about "defining technology."

B. New and Modified Categories and Classifications

As a prologue to a systematic discussion of the different aspects of what constitutes a technology, it is useful to consider some examples of changes in the classes of things that are recognized in common usage and in science. Prior to their arrival in North America, "rattlesnake" would have represented no meaningful category to the English. However, that changed following their encounter with venomous snakes that made a rattling or buzzing sound with their tails. Before Evangelista Torricelli (1608–47) constructed what came to be recognized as the first barometer around 1643, few believed that the air had weight; moreover, probably no one gave any thought to the possibility that changes in the height of the column of mercury in a device like Torricelli's would correlate with changes in the weather. Soon after Torricelli's results became known, others who constructed similar devices noticed that a drop in the height of the column of mercury correlated with cloudy and/or rainy weather and an increase in its height with fairer weather. Thus did the concept of atmospheric pressure and its meteorological significance take their places in human thought.

In the fourteenth century, a group of mathematicians and logicians, dissatisfied with imprecise discussions of motion and speed, set about giving a rigorous treatment of motion that is changing (e.g., speeding up or slowing down). In the process, they distinguished uniformly or constantly accelerated motion from motions of other types (whether uniform or changing in less regular ways) and in doing so laid the foundation for Isaac Newton's (1643–1727) work, which stands as the culmination of the Great Scientific Revolution of the sixteenth and seventeenth centuries. (Modern readers accustomed to speedometer reading may find it surprising that the concept of instantaneous velocity poses problems. An attempt to formulate a clear and precise method of determining the slope of a curved line at a point should dispel this illusion.)

Carolus Linnaeus (1707–78, also known as Karl von Linné) steered the taxonomy of plants in a fruitful new direction when he set out to classify them on the basis of the structure of their reproductive parts (flowers) instead of such features as habit of growth (tree or shrub) and seasonal cycle (evergreen or deciduous). Linnaeus was, of course, developing a new classification system, not simply introducing a single new category, but he did put forward new standards for inclusion in the classes, subclasses, etc. of plants in his system. Quite a variety of natural occurrences, not previously thought to be related (amber's ability to attract tiny bits of straw, Saint Elmo's fire, lightning, etc.) were pulled together into a new class, "electrical phenomena," as scientists explored this mysterious new realm. The basis for inclusion in this new class was being a manifestation of electricity.

Changes in classifications and the bases for them can be illustrated by the different ways whales have been treated. Despite the widespread (and persistent) tendency to lump whales together with fish, Aristotle placed them and dolphins in their own category, separate from fish, birds, oviparous quadrupeds (reptiles and amphibians), and viviparous quadrupeds (roughly what are now classified as mammals). Aristotle based his classification of animals on their bodily structure, their physical capabilities, their ways of sustaining and reproducing themselves, and their behavior. Subsequently, the quadruped criterion was dropped; more emphasis was placed upon the basic life processes that whales share with warm-blooded land animals, including the young being nourished by their mother's milk; and whales were included in the (mammal) classification. However, this does not mean that the category "quadruped" is somehow defective or that membership in it fails to be based on shared characteristics. Rather, as in the case of Linnaeus, one is faced here with changes in the factors considered in classifying (for certain purposes) life forms and in the emphasis placed upon each of them.

And even more fundamental changes were to follow. With the acceptance (contrary to Aristotle) of the idea that terrestrial life forms are not static but evolve over time so that one ancient species may be the ancestor of a number of distinct modern ones, the goal of taxonomists became to place each organism in the correct location on the evolutionary family tree, irrespective of presence or absence of this or that similarity in existing life forms.

These changes in classificatory criteria also affected the makeup of the whale class itself. Although huge or impressive size—as in "whale shark" and "a whale of a sale"—has always been one of the things conveyed by "whale," biologists' recognition of systemic similarities between sperm whales (which may reach 60 feet in length), beaked whales (up to 40 feet), killer whales (up to 30 feet), pilot whales (up to 28 feet), belugas and narwhals (from 10 to 15 feet), dolphins (from 7 to 13 feet, depending on the species), and porpoises (which may reach only about 5 feet) led to the inclusion of the smaller of these marine mammals in the same suborder as the larger ones. Given the name "odontoceti" (New Latin for "toothed whales"), it is a suborder of the "cetaceans," a more precisely defined group than "whales" that includes (in addition to the odontoceti) the baleen whales, which have no teeth. Although some cetaceans are enormous, others are not. Nevertheless, for certain purposes (e.g., in discussing the history of the whaling industry) it is useful to be able to refer to the class of extremely large marine mammals superficially resembling fish as "whales," irrespective of whether or not some of them are more closely related biologically to the smaller odontoceti (which are not counted as whales in this sense) than they are to the rest of these extremely large marine mammals. Furthermore, one must not assume that this latter, more traditional usage is somehow incorrect or simply too crude to be useful. Rather, the most important point is that the two usages serve different ends. The designation "cetacean" has an explicit taxonomic and explanatory as well as a descriptive function, whereas the traditional usage has a practical and a (somewhat different) descriptive one.

Evidently, then, things can be categorized in different ways to serve different purposes, and new ways of categorizing things may accompany and/or help to bring about advances in human understanding. Bringing together under the heading "technology" human activities which seem to have certain characteristics in common and which have had and continue to have a major impact upon human life and the environment is just such an undertaking. However, because of the wide variety of views on what should be included in this category, formulating a usable definition of "technology" will require making decisions about a number of aspects of the concept. In making these decisions, it will be important to keep in mind the end(s) to be served by the category "technology" and to consider the implications of particular ways of drawing its boundaries.

C. Basic Aspects of Technology

Despite the divergent overall views regarding how technology is to be understood, there are some aspects of it that are not in dispute. First of all, technologies are not just tools, machines, electrical systems, etc.; they are complexes that may include one or another of these physical objects or systems, but they also involve an entire set of **activities and practices**. For example, what makes a screwdriver a screwdriver is not just its shape and physical make-up, but what people do with it. Rather than an isolated, spur-of-the-moment action, driving or removing screws takes place in the context of an entire constellation of devices and techniques brought into being and maintained within a cultural matrix. Mining, extracting metals from their ores, and fashioning screwdrivers (and screws) from metal and other materials involves many skills, instilled by long practice, and complex processes unimaginable except in a human society with enduring traditions.

A second basic feature of technology concerning which there is little disagreement is its **utilitarian** nature. Technology is one of those things that is "good for" something else. People develop and use technologies to accomplish things that would be more laborious, more difficult, or impossible without them. Technology is an activity that is engaged in not for its own sake alone but to obtain a desired object or to bring about a desired state of affairs. But is this its only identifying feature? Should the category include *every* activity pursued in order to obtain or bring about something else?

Unfortunately, there had been no agreement on this issue among philosophers, social critics, and other thinkers. Accordingly, it is necessary to make some decisions regarding just which kinds of activities are to be included under the heading "technology," and reasons for them must be given based upon the purposes of such a classification and upon the consequences of setting its boundaries in a certain way. Consider, first of all, the question of whether activities in which the only resources utilized are those of the human body alone, for example, climbing a tree and picking an apple in order to eat it. There are at least three reasons not to include, under the heading "technology," activities like the one just cited.

First, such a move would render the introduction of the term "technology" pointless, since there already exist ways of identifying activities that are undertaken for the sake of something else. Second, such a classification would encompass human activities for which there are clear parallels in much animal behavior. This raises the question of whether animal behavior of this kind should be included along with human activities under the rubric "technology." The absence of obvious reasons to exclude such animal behavior, if technology is broadly defined in the way under discussion, makes it difficult to see how technology could play any significant role in accounting for the emergence of those characteristics that set humankind apart from even its closest biological relatives. These themes will be explored in the next as well as subsequent chapters. Similarly, and this is the third point, something much more specific than "actions undertaken for the sake of something else" is required to account for the appearance at the time of the Industrial Revolution of the social and environmental ills that captured the attention of social critics, philosophers, and other thinkers.

Thus, it would appear reasonable to accept as a third criterion for something to be classified as a technology the requirement that it include a **material component** in addition to the bodies of those utilizing it. This means that technology involves the *use* of physical objects or natural processes to accomplish things that would be more difficult, more time consuming, or simply impossible without them. Simple examples would include using sticks to defend oneself or to attack prey, rocks to crack open nuts or bones (to gain access to the marrow), sharp-edged broken rocks to cut things, broken gourds to dip water, etc. Activities of this kind or similar ones are, of course, not confined to human beings but have been observed among animals. For example, sea otters, while swimming on their backs, sometimes balance a stone on their bellies and hammer shellfish against it in order to break them open. Chimpanzees and orangutans insert sticks or grass stems into termite nests and devour the insects clinging to them when they are pulled out. Consequently, the *use of tools* alone cannot be said to capture the essence of technology for two of the same reasons (the second and third) given above for concluding that the utilitarian nature of technology was insufficient to fix its identity. It should be noted that recognizing this third feature as characteristic of technology also distinguishes such technological fields as engineering, metallurgy, electronics, and medical technology from rhetoric, public speaking, the exercise of leadership, politics, pure mathematics, etc. More will be said on this topic in subsequent chapters.

If the use of tools is not sufficient to qualify an activity as technological in nature, then what more is required? It will doubtlessly occur right away to the astute reader that, as has long been recognized by anthropologists, the answer is **tool making**, which has played an essential role in the (biological) success of the human species. Without it, the deplorable conditions condemned by social critics and others would never have arisen. It thus appears reasonable to recognize as a fourth feature of technology that it involves making tools as well as using them. Here it must be stressed that, for present purposes, "tool" is to be understood broadly as an instrument, a means of accomplishing something. "Tool-making" thus encompasses a wide range of activities, from the production of the simplest kitchen knife to the design and construction of hydroelectric dams and space shuttles. If the previously identified characteristics of technology are taken into account, this fourth feature amounts to the claim that technology involves altering physical objects or systems or the course of development of physical, chemical, or biological processes in such a way that they better serve practical purposes than they would if left in their unaltered conditions.

The question at this point is whether the criteria set out thus far are sufficient. That is, are there activities that would count as technological in nature, based on the criteria enumerated so far, but which really should not be so considered? Are there activities which, for good reasons, should not be classified as technological in nature even though they effect changes (in physical objects or systems or the course of development of physical, chemical, or biological processes) which provide no direct or immediate benefits but do make possible or more effective other activities that do provide such benefits? If so, the question to be addressed is what additional characteristic(s) of technology need to be identified and incorporated into a usable definition of it.

In addressing this question, it is instructive to consider the case of beaver dams. Like hydroelectric dams, they provide benefits only indirectly: the pond that forms behind one of these dams provides a suitable location for beaver lodges with their underwater entrances and for the beavers' winter food supply: an underwater stash of logs and branches, the bark and twigs of which beavers eat. As enshrined in the phrase "busy as a beaver," beavers seem always to be working: cutting down trees, removing limbs, and using the logs and branches to build, repair, or add to their dams or lodges or to build up their food supply. While these activities are impressive, there are good reasons to believe that they are instinctive. Beavers assiduously cut down trees even in the absence of a place to build a dam or a lodge and even when their food supply is more than adequate.

By their very nature, instinctive behavioral patterns remain the same from generation to generation. There simply is no place for innovation except for the glacial process of genetic variation and natural or artificial selection. In contrast, by the time of the Industrial Revolution people could see a way of life being swept away by technological innovation right before their eyes. Nowadays many electronic devices are on their way to becoming obsolete by the time they leave the factory. Technology is not static. Indeed, it is technological innovation—piled upon technological innovation—that makes technology so powerful and so important, so beneficial yet so destructive. It follows that behavior that is instinctive or "hard-wired" does not belong in the class of technological activities. This means that some activities that animals engage in which might appear to belong in the technology category really do not belong there at all. Thus no matter how impressive is the intricate architecture of spider webs, the hexagonal wax cells of bees, the dangling globular nests of weaverbirds, and the dams and lodges of beavers, they do not belong in the same category as suspension bridges, gothic cathedrals, and aqueducts, not because the latter are so much larger and more complex but because they exemplify technological innovation in contrast to the repetition of behavioral patterns from generation to generation to generation.

As is surely evident from a review of the history of technology and especially from a study of the origins of those inventions that bulk so large in contemporary life, **innovation** and the **potential for innovation** are intrinsic aspects of technology. This then is the fifth fundamental characteristic of technology which must be incorporated in any definition of it. Note in particular the "snowball effect" described in the Introduction. All of these aspects of technology will return throughout the book, but of particular importance are innovation and the incremental, gradual changes that sometimes result in very slow development and occasionally allow for massive leaps (or exponential growth).

In general terms, then, technology (i) consists of activities and practices which (ii) are engaged in to achieve or promote something else (iii) by forming and controlling (iv) material objects and processes ("tool making" and "tool using") and which (v) are not instinctive but by their very nature involve innovation and the potential for innovation.

D. Complicating Factors

Even so, in many societies technological innovation has been minimal or absent over extended periods of time. More conspicuously absent is an awareness of technological innovation. Technology itself is, of course, an important part of every human culture, but in many societies it is unlikely to be understood in a way that is at all like that developed in this work. Rather, until relatively recently and in some places even now, the basic technological practices of a society tend to be viewed as a given, and any account of their origins is likely to take the form of narratives recounting the exploits of powerful legendary beings in times long past, as exemplified by the story of Prometheus, who according to Greek mythology stole fire from Olympus and bestowed it upon humankind. Alternatively, some may believe that powerful spirits may be actually present in technological activities and processes, as traditionally was the case in the making of the great wooden canoes of the Native Americans of the Pacific Northwest (discussed in detail in Chapter IV).

Sometimes, however, new crafts emerge in societies in which traditional ways of doing things predominate. There is compelling evidence that the female basket-makers of the Pacific Northwest initiated a new use for their basket-making techniques: weaving blankets and robes. There could hardly be a better example of a new technology being developed from an existing one (a topic also covered in Chapter IV). Indeed, the technology of the people of the Pacific Northwest is an excellent example of the "snowball effect" noted above. Even though they did not practice agriculture, these Native Americans were able, because of a special set of circumstances, to enjoy a sedentary or semi-sedentary way of life. In a pattern repeated around the world and across many thousands of years, their dwellings also served as workshops and warehouses supporting a rich repertoire of technology activities (also described in detail in Chapter IV).

One must remember that existing technological resources (the "snowball effect") is but one of the factors affecting the rate at which technological innovation takes place and, indeed, whether it happens at all. Technological activities, including technological innovation, always take place within a social and cultural matrix, and such factors heavily influence human behavior, including innovation of all kinds. Accordingly, there has been much interest in identifying and understanding those factors that bear upon technological innovation. Since technology affects human life in so many ways (some of them noted in the Introduction), the relationship between technology and other aspects of the human scene is very complex. Thus, it is not surprising that so much of what has been said or written about technology addresses such matters, as does much of the material in this book. Chapter VIII is focused on this topic and includes a case study of the very different roles played by gunpowder in three cultures.

Of course, the greatest complicating factor, suggested earlier in this chapter and to which we now turn, is the relationship between technology and science. Although technology and science are ultimately distinct, the boundary between the two is complicated, as the discussion and case study below demonstrate.

E. Science and Technology

Not only is technology much older than science (which entered the picture only as literacy blossomed), but science and technology have not always marched along together and been as mutually supportive as they are currently. Indeed, one or the other has flourished at times and places while the other did little more than escape stagnation. For example, the Ancient Greeks made significant advances in a number of scientific fields, but their technological innovations (though not unimportant) were rather modest in comparison and did not, to any significant degree, depend upon or derive from their scientific accomplishments. Conversely, the Romans excelled in engineering but did little to advance science. The scientific achievements of the Europeans during the Late Middle Ages were also quite modest, but technological advances accelerated to the extent that historians sometimes speak of the "Industrial Revolution of the Middle Ages." During this time, new technologies such as the windmill, printing press, and cannons changed the way people harnessed energy to do work, disseminated information, and conducted warfare. Certainly, as will be seen in the following case study, some of the core technologies of the Industrial Revolution could not have played their robust roles in this historic transformation of human society without the innovations made during the Middle Ages. In any case, in neither of the two periods did science play a direct, major role in the development of new technologies—even though the Industrial Revolution was preceded by the Great Scientific Revolution initiated by Copernicus (1473–1543) and culminating in Newton. Furthermore, studies carried out in recent decades have found that most new technologies continue to derive from existing devices and techniques rather than directly from the application of scientific advances *per se*.

Although technology continues to have an identity separate from and independent of science, science has been a central factor in the development of technology during roughly the last century and a half. The two are closely related currently. Consequently, clarifying the nature of the relationship between the two must be a central part of any serious discussion of the nature of technology. The essential part played by technology in the actual practice of modern science is forcefully demonstrated by what happens when powerful new technologies usher in new procedures for accomplishing the basic tasks that scientists routinely need to accomplish. Such a transition is epitomized by what has been called "The Instrumental Revolution in Chemistry," which is the focus of the case study that closes this chapter.

1. A Porous and Diffuse Boundary

One might think that, prior to discussing how science and technology relate to each other, one would need to address the question of just what is to be included under the rubric "science," just as was done in section C above regarding "technology." And that would be correct if a systematic treatment of the subject were contemplated. But since the purpose of this section is to provide only an introduction to the topic, it suffices, firstly, to say that the concern here is with science in the first of the three senses considered near the beginning of section A (i.e., an ongoing tradition of inquiry seeking knowledge and understanding of the natural order) and not with science in any other sense of the term. Secondly, care will be taken to ensure that examples used are drawn from disciplines whose status as sciences is not problematic, such as astronomy, physics, chemistry, biology, geology, etc. As simple as it is, this characterization of science provides a useful starting point for a discussion of how scientific and technological activities relate to each other.

To begin with, it should be kept in mind that the difference between the two can be rather subtle. Whether a device or process is a scientific instrument or procedure rather than something employed in practical affairs may depend on nothing more than the use to which it is put. There could hardly be a better example than Galileo's (1564–1642) telescope, which Galileo initially tried to market as a device useful to naval forces (which it certainly became). But when an opportunity arose for him to capitalize on its possibilities as an astronomical instrument, he lost no time in pursuing it—with revolutionary results. It should be noted that Galileo's personal motives for engaging in scientific endeavors—be they "pure" (the pursuit of truth) or self-promoting or enriching—are irrelevant to their scientific status and merits.

A more recent, powerful example of an invention ricocheting between scientific and non-scientific uses is afforded by modern rocketry. Robert Goddard (1882–1945) launched the discipline almost single-handedly as a vehicle with which to explore outer space, a scientific undertaking. Almost immediately Goddard's results were enlisted in the German armaments buildup during the Second World War, resulting in V-2 ballistic missiles descending upon the cities of the Allies. In a strange twist of irony, Wernher von Braun, a key player in the German V-2 program, later led a team that developed rockets for the U.S. military and the U.S. space program, including work on the Saturn rockets, one of which launched Neil Armstrong on his historic trip to the moon.

Hardly anyone would deny that Goddard was doing scientific work or that he was a scientist. After all, he was a physics professor who ended up devoting his life to developing a device that would vastly expand human knowledge of the physical universe. But what of the work done by others who wrestled with guidance and propulsion-system problems not unlike those that occupied Goddard but who did so in the V-2 program or in one of the U.S. military rocket programs? (Of course, the latter eventually did provide launch vehicles for space exploration.) Given the similarity, on the concrete level, of their work to Goddard's, and given such commonly used expressions as "It doesn't take a rocket scientist to ...," one might be inclined to answer in the affirmative, "Yes, they are doing scientific work, and they are scientists too." On the other hand, does it really make sense to count as a "scientist" anyone who invents or helps develop a device that later is used for scientific purposes? Furthermore, "rocket scientists" do not study rockets in the same sense that botanists study plants. Rather, they design rockets and endeavor to get them to work, which surely is a technological undertaking. Of course, members of rocket teams may have been educated as scientists, may have done scientific research, and may use scientific knowledge in their work, but their immediate objective need be no more than to bring about a desired state of affairs.

Under what conditions, then, should those who are employed in technological capacities in support of scientific projects be considered scientists doing scientific work? Minimally, it would seem reasonable to say that they must have some understanding of and interest in the scientific purpose of these projects. If this condition is not met, there would seem to be no more reason to include them in the category "scientists" than there is to include machinists, electronics technicians, and others who make or refine scientific instruments according to specifications given to them by scientists. Even so, it must be kept in mind that contributions to the technological aspects of science, whether or not they are made by scientists, are absolutely essential to it. And, of course, scientists who do make such contributions are themselves engaging in technological activities.

Furthermore, it should be recognized that a person can engage in both practical and scientific pursuits simultaneously, perhaps by working on the same project or closely related ones, and that a person can shift the focus of his or her efforts from one of these areas to the other without changing very much of what he or she does on a concrete level. Thus, the boundary between science and technology is not only porous but also diffuse. Among those inhabiting this nebulous border region are persons who received a scientific education and are employed, based on their knowledge and skills, by industries, government agencies, foundations, etc. to perform tasks that may or may not advance the goals of science. Additional illustrations of this point are found below.

2. The Technology of Science

Several different types of connections exist between science and technology. The most widely recognized type involves utilizing scientific knowledge in solving technological problems, including the invention of new devices and techniques. The general acceptance of the view of that technology as "applied science" stems from the myriad achievements that have come about in this way; merely cataloging them would be a major undertaking. Key examples discussed later include Watt's drawing upon available knowledge of elastic fluids to reach a solution to problems plaguing the Newcomen steam engine and the exploitation of Faraday's discoveries concerning electromagnetism in the development of electric motors. However, since these episodes are also profoundly connected to the relationship between economic factors and technology and science, these cases will be considered later in Chapter VII (Technology, Economics, and Science).

An equally important connection is the indispensable role played by the technological resources of science in its development. Its significance is evident just from noting the devices deployed in launching modern science, among them Tycho Brahe's (1546–1601) giant quadrants, Galileo's telescope, Galileo's bronze sphere rolling on a sloping, grooved track, Torricelli's barometer, and Leeuwenhoek's (1632–1723) microscope. Sophisticated technologies were crucial to the invention of such instruments. Lens-grinding for spectacles, initiated in northern Italy in the thirteenth century, paved the way for Galileo's telescope. The glass tubes made by Torricelli and co-workers in 1643 reflect the advanced state of the glass-making art in mid-seventeenth century Italy. They probably could not have been made anywhere else in Europe at the time. Like other notable astronomers and experimental scientists, those referred to above made use of existing technologies, but their forte was building on, modifying, or incorporating them into new instrumentalities. This feature of scientific practice has important implications for the way science develops, for the way new technologies find their way into general use, and for the relevance of economic factors to science (and vice versa)—implications that have not always been recognized and rarely given the attention they deserve.

To begin with, new devices constructed for a given scientific purpose may reveal phenomena not previously detectable or even suspected and/or make it possible (or economically feasible) to explore phenomena and relations among them that otherwise would not have been investigated. Thus, the invention of a scientific instrument may have a profound influence upon the direction of scientific research as well as provide a better way to address practical problems. A good example is the barometer. This invention seems to have sprung from

Evangelista Torricelli's efforts to demonstrate two things: (a) the correctness of his proposed explanation of the inability of suction pumps to draw water (vertically) more than some 30 to 34 feet and (b) the possibility of a completely empty space, a vacuum. Torricelli maintained that air has weight and that it is atmospheric pressure that pushes water up pipes attached to suction pumps. In his view, such a device simply serves to remove air from the attached pipe, reducing the air pressure in it. Atmospheric pressure on the water in the reservoir in which the lower end of the pipe is submerged then pushes water up the pipe—but only to the height at which it is balanced by the pressure deriving from the weight of the water in the pipe. As Torricelli saw it, when an attempt is made to pump water more than 34 feet vertically (with a suction pump), the most that can be achieved is a vacuum in the space in the pipe above that level. At the time, this was an especially audacious suggestion for two reasons: first, the long established view, attributed to Aristotle, was that a vacuum could not exist ("Nature abhors a vacuum"); second, Aristotle's system of the "four elements" (earth, water, air, fire) had not yet been discredited. According to it, all things in the natural world consist of combinations of these four. In addition to a fluid nature, air was believed to possess "levity," the very opposite of weight, as seen in its bubbling up when released under water. At the time, the identity of the chemical elements and the difference between a chemical reaction and a change of phase (melting or evaporating, for example) had not yet been learned.

Since he had been an assistant to Galileo, it is no surprise that Torricelli was willing to entertain unconventional ideas about the natural order. Recognizing that, according to his proposed explanation, suction pumps would be even less effective with liquids that are denser (of greater weight per unit volume), Torricelli turned his attention to mercury. He reasoned that since mercury is 14 times as dense as water, air pressure would support a column of mercury under similar circumstances only one fourteenth as high (or about 29 inches). This view was vindicated when he (and possibly others who had been assistants to Galileo) took a glass tube, about three feet in length and sealed at one end, and filled it with mercury. He then covered the open end with his finger and submerged that end of the tube in a dish of mercury. When he removed his finger and raised the tube to a vertical position, the mercury drained out of the tube into the dish until the height of the column of mercury in the tube stood at about 29 inches, leaving an empty space at the top (see Figure 2-1). Torricelli argued that since no air could have gotten into the tube, the space left at the top of the tube must be a vacuum.

Of course, this result did not prove definitively that Torricelli was correct on either count. Long standing, deeply rooted beliefs about the natural order are

Figure 2-1. Torricelli and the Barometer. Unknown artist. CC-BY-4.0.

not easily dethroned, more so then than now. After all, it was not possible to tell by visual inspection that the "empty" space at the top of the tube was a vacuum. Furthermore, other explanations for the observed phenomena were possible, as was pointed out at the time. But additional evidence was soon obtained by French philosopher Blaise Pascal (1623–62). He reasoned that if Torricelli's experiment were carried out on the top of a high mountain, the height of the column of mercury that resulted would be shorter than had been observed earlier (since there would be less air above, pressing down on the mercury in the dish). Pascal arranged for his brother-in-law, Florin Périer (1605–72), to carry out just such an experiment on the Puy-de-Dome in southern France in 1648. As a control, Périer left an assistant with an apparatus like that used by Torricelli at the base of the 4,806-foot mountain while he ascended it with another such apparatus. He found that the column of mercury was indeed a few inches shorter at the summit. Subsequent experiments yielded results consistent with those obtained by Torricelli and Pascal. The implication of these results is that if Torricelli's experiment were carried out at higher and higher elevations, the resulting heights of the column of mercury would be shorter and shorter until the level of the mercury in the tube was the same as that in the dish, i.e., no air pressure at all, just empty space at that elevation.

This empirically-based conclusion can be established simply by considering that if the air extended outward without limit, so would the air pressure increase, crushing organisms like us. The discovery that the Earth is surrounded by an ocean of air with empty space beyond was but the first contribution made by means of Torricelli's instrument to the emerging scientific picture of the home of the human race. It was followed almost immediately by another: the connection between barometric pressure and meteorological conditions. These two discoveries had a more direct and significant bearing on people's daily lives than did the heliocentrism of Copernicus.

Torricelli and others noticed immediately that the height of the column of mercury in devices like his was not constant but varied with the weather: greater heights (greater atmospheric pressure) correlated with fair weather, lower heights (lower atmospheric pressure) correlated with precipitation. A whole new era in the study of the weather began as scientists set about recording and studying correlations between weather patterns and barometric data from different localities. Only four years after Torricelli's successful experiment, the great French mathematician and thinker René Descartes was writing (from Holland) detailed instructions to Marin Mersenne (in Paris) about keeping records of barometric readings for purposes of comparison. Mersenne, who engaged in a voluminous correspondence with the luminaries of the time, is credited with spreading the word about Torricelli's achievements. The barometer quickly became a standard item in the tool bag of scientists working to identify, classify, and understand meteorological phenomena.

That Torricelli is most often simply credited with inventing the barometer (credit which is well deserved—but by no means the whole story) illustrates how easy it is for people to lose sight of the role played by technology in determining the direction of scientific investigation. Given the longstanding recognition of atmospheric pressure and of the connection between it and weather patterns, it is hardly surprising that people assume—absent specific knowledge to the contrary—that the motivation for making an instrument so evidently capable of registering changes in atmospheric pressure was to produce an instrument that would register such changes. What this assumption ignores is that, in the absence of such an instrument, no one would have known that there were such changes in atmospheric pressure or, indeed, that there really was such a thing as atmospheric pressure at all (although some thinkers did subscribe to theories that have that implication).

Just as rapidly as it was pressed into the service of science, the barometer was put to practical use in efforts to provide people with valuable information. Falling barometric pressure gives forewarning of storms. High (and stable) barometric pressure signals continued fair weather. In 1646, Otto von Guericke, who had succeeded (after some ten years of effort) in producing a vacuum by means of a vacuum pump, was elected burgermeister of Magdeburg. In that capacity, he arranged for the construction of a water barometer over 34 feet tall attached to the side of a house where it could be seen by passersby. Consisting of a brass tube with a closed glass section at the top, it featured a figure of a man that floated high in fair weather and sank when it was stormy. From its beginnings, meteorology has been both an important practical art as well as a field of scientific investigation.

Another leading seventeenth century scientist who could have performed none of the experiments for which he is remembered without previous technological resources is Robert Boyle (1627–91). As had Torricelli, Boyle took advantage of the advanced glass-working methods of the time and made use of glass tubes sealed at one end. With tubes made of glass, it is possible to see what is happening as an experiment progresses: e.g., whether the mercury level in a tube has fallen (or risen) and by how much. Boyle employed U-shaped glass tubes resembling inverted staples or tuning forks in a vertical position. To begin one of his experiments, he or his assistant(s) poured mercury into the open end of one of these tubes until the curved section at the bottom was full enough for the tube to be completely blocked. Doing so produced an enclosed volume of gas (air) that could be subjected to increases in pressure simply by adding mercury to the open end of the U-shaped tube. It was immediately evident that an increase in air pressure in the closed part of the tube resulted in a decrease in its volume. Careful measurements revealed that, for example, doubling the pressure reduces the volume of the gas by half. In general, he found that the product of the pressure and the volume remains constant during such changes (at fixed temperatures), a relationship that became known as "Boyle's Law." Also essential to Boyle's work were large glass dome-like objects ("bell jars") which, when used in conjunction with vacuum pumps (also recently perfected, as discussed above and seen below in Figure 2-2), enabled him to study such things as the melting and boiling points of liquids.

There could hardly be a clearer demonstration of the heavy dependence of a scientific discipline upon its technology or of the practical utilization thereof than the above episode from the history of meteorology. But the extent to which the practice of science involves carrying out technological operations can be seen by examining not only (a) the early history of one or another science but also (b) cases in which a radically new technology has transformed what the practitioners of a given science actually do and, as a result, how they think about what they are doing. An example of the practice of a scientific discipline thus being fundamentally changed is afforded by developments in chemistry in the latter half of the twentieth century. During this period, the techniques of "wet chemistry" were replaced by procedures employing electronic instrumentation. That is the topic of the following case study.

3. Case Study: The Instrumental Revolution in Chemistry

Until roughly the middle of the twentieth century, analytical chemists seeking to separate and identify the components of complex mixtures would proceed by applying solvents and reagents (often acid solutions), observing color changes in solutions, noting the conditions under which precipitants were formed, etc. In contrast, a chemist currently seeking to identify the components of a mixture is likely to employ a technique known as chromatography to separate the components of the mixture and to use various electronic devices to pin down their identities. In the vast majority of cases, it is necessary to separate the components of a complex mixture before attempting to identify them because the presence of a second (or third or ...) substance may interfere with tests or procedures being carried out with regard to the first. A few years before the beginning of the Second World War, it began to be recognized that the best method of effecting this separation was chromatography.

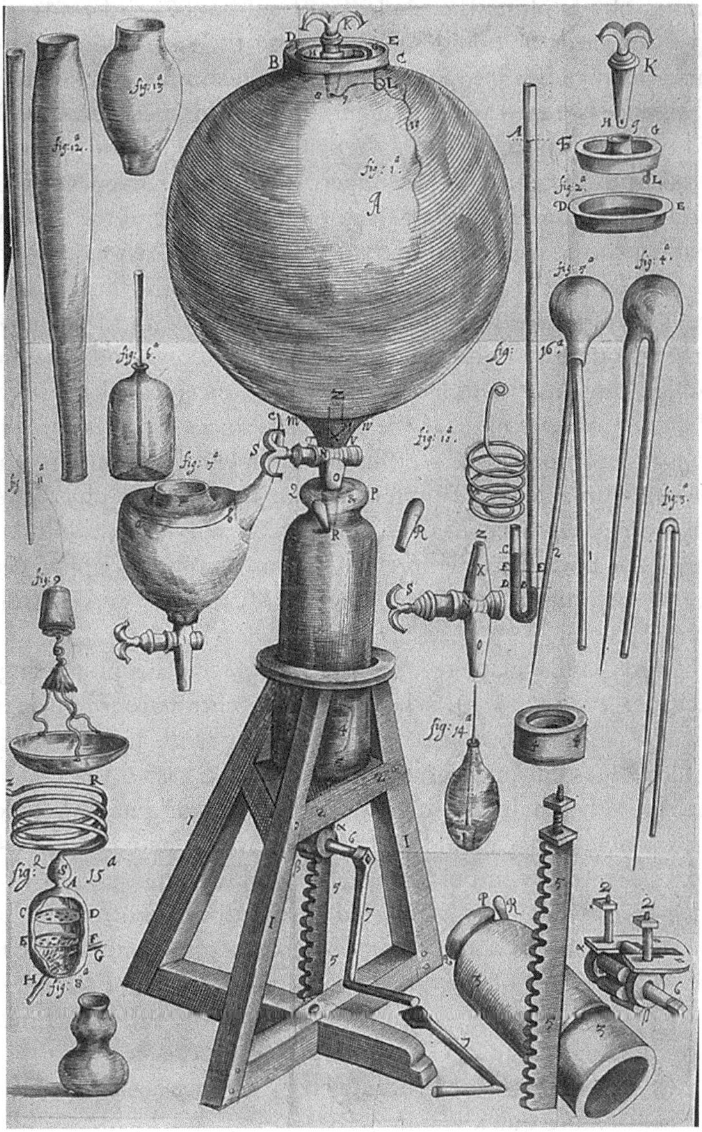

Figure 2-2. Robert Boyle's Vacuum Pump. This is an illustration from Boyle's 1660 work *New Experiments Physico-Mechanicall, Touching the Spring of the Air, and Its Effects (Made, for the Most Part, in a New Pneumatical Engine)*, depicting one of his designs for a pump that allows a user to turn the hand-crank at the bottom to remove air from the glass globe on top, creating a vacuum.

Chromatography exploits the tendency of substances to be adsorbed at different rates from gases or liquid solutions onto the surfaces of solids or to be dissolved at different rates into liquids. Thus, for example, if a solution containing two substances (solutes) with different affinities for a certain solid material is poured into the uppermost part of a vertical tube packed with particles of that solid material and (except for that uppermost section) filled with the solvent involved, and if the liquid is allowed to exit at the bottom of the tube while more solvent is added to the top, then the two solutes will be washed down through the column of solid particles by the flow of the solvent. But since the two solutes have different affinities for the solid material, the one with the greater tendency to be adsorbed to the solid particles will be washed down the tube more slowly than the other. Thus, as time passes, it will be concentrated nearer the top of the tube than the other solute. If the two substances (solutes) are different in color, their concentrations at two different levels will be manifested as two differently colored bands in the column of liquid. To effect the physical separation of the solutes, one simply has to capture the fluid carrying one solute before the other arrives at the bottom of the tube. The term "chromatography" derives from the presence of such differently colored bands. ("Chroma" derives from a Greek word meaning "color.")

These colored bands played a prominent role in the pioneering work of Mikhail S. Tsvet (1872–1919), who developed chromatography early in the twentieth century as a way of separating plant pigments. In an alternative form of this technique (gas chromatography), an inert gas, rather than a solvent, carries the material to be analyzed (in vapor form) through a narrow tube in which it comes into contact with a solid or liquid for which its constituents have differing affinities. Especially important is gas-liquid chromatography (or just "gas chromatography" as it is usually referred to). In this approach, the gas passing through the tube comes in contact with a liquid kept in place by being bonded either to the inner surface of the tube or to the surfaces of the solid particles filling it or lining it. In this case, the method exploits the differing rates at which gases dissolve in a given liquid. Gases that dissolve more slowly in the fluid they encounter in the tube are carried through it by the inert gas more rapidly than those that dissolve more quickly in it.

The presence of concentrated amounts of solutes in the liquid exiting a chromatographic tube can be registered by a detector and displayed on a graph, with the strength of the signal (corresponding to the concentration of the solute) plotted along one axis and with time plotted along the other. Such a display is called a chromatogram. Similarly, in gas chromatography, the presence of concentrations of vapors in the carrier gas exiting the chromatographic tube can be detected and displayed in the form of a chromatogram. A (possible) first step in identifying the components of a mixture is to compare the chromatogram obtained from it with those obtained from known substances. Indeed, in contemporary installations, chromatographic detectors are likely to be connected to computers, making it possible to draw upon extensive databases for purposes of comparison and analysis. The usefulness of such setups is attested by the presence (by the mid-1980s) of the many thousands of chromatographs in petrochemical complexes. However, in other than special circumstances, additional data must be gathered to effect positive identifications.

While quite a variety of methods can be used to identify the constituents of a mixture once they have been separated by chromatography, the mass spectrometer has special advantages. It is effective even with very tiny amounts of material, can provide both quantitative and qualitative information about substances, and can distinguish substances even when one does not lag behind the other in passing through the tube of a gas chromatograph. The mass spectrometer is based upon the principle that an electrically charged body moving at a right angle to a magnetic field is subjected to a force proportional to its charge and velocity but perpendicular to its direction of motion and perpendicular to the direction of the magnetic field. Thus, a charged particle, an ion for example, moving between the poles of a magnet will be deflected in a curved path; and a less massive particle will be deflected more than a heavier one of the same charge and velocity (since they are subjected to the same force but have unequal masses). The mass spectrometer makes use of this difference to distinguish one type of particle from another on the basis of the difference between their masses. Its first triumph was the confirmation, early in the twentieth century, of the existence of isotopes, species of substances that are chemically identical but made up of atoms of different masses. However, it was only in the second half of the twentieth century that the instrument's full potential for chemistry began to be realized.

By using a high-resolution mass spectrometer to aid in identifying the fragments of complex organic molecules, chemists rapidly made major strides in working out the structures of these molecules. As a result of such technological innovations, structural studies that might have been the focus of an entire scientific career became projects for graduate students working with technicians. While several chemists played a role in bringing about this transformation, the contributions of Klaus Biemann (1926–2016) of MIT are especially worthy of note. After pioneering work with the mass spectrometer unraveling the mysteries of organic compounds, Biemann focused his attention in the mid-1960s on combining mass spectroscopy with gas chromatography. Because of its ability to obtain results from samples as small as a few micrograms, the mass spectrometer is ideally suited for use in identifying the minute quantities of pure substances isolated by gas chromatography. By 1975 the gas chromatograph/mass spectroscope was sufficiently well developed for the instrument to be utilized in the Viking mission to Mars. Following its successful completion, the team in charge of this part of the mission (led by Biemann) concluded that there are no organic compounds on Mars, a very surprising—and, to many people, very disappointing—finding.

Despite their remarkable capabilities, the size, weight, and cost of mass spectrometers utilizing magnets limits their usefulness. Accordingly, scientists focused their efforts on developing alternative forms of the instrument. Perhaps most significant is a device known as the quadrupole mass spectrometer or mass analyzer. Instead of a magnet, it features four parallel, electrically charged metal rods configured like the legs of a square table but positioned much closer together. A constant electrical voltage with a rapidly oscillating voltage superimposed upon it is applied to the metal rods. The electric field thus produced in the space between the rods can be adjusted so that a positively charged ion introduced into one end of that space will travel the entire length thereof (rather than wandering off to one side, colliding with one of the rods) only if its mass has a certain value. In that case, its arrival at the other end will be registered by a detector and recorded.

A major contributor to the transformation of the practice of chemistry was the computer revolution, especially the availability of relatively small and inexpensive units. These made it possible to record massive amounts of data and to manipulate it with remarkable speed. Furthermore, computer control of the quadrupole mass spectrometer makes it possible to test for the presence of particles of an entire range of masses in an incredibly short period of time by rapidly varying the voltages applied to the metal rods of the device. Placing it and the gas chromatograph under the control of a single computer enables even more sophisticated operations. With no heavy magnets and a greatly reduced price tag, the quadrupole gas chromatograph/mass spectrometer quickly demonstrated its capabilities and was put to work not only in research efforts and industrial activities but also in monitoring environmental conditions, checking the quality of food products, etc. More recently these devices have been used in security procedures. TSA (Transportation Security Administration) agents at airports place the cotton swabs that they run over passengers' luggage into these devices to test for the presence of explosives.

We'll return to the cases of the gas chromatograph and mass spectrometer in Chapter VII, when discussing in greater depth the relationship between technology, science, and economics. Moreover, we'll examine the transformative impact of computers and the digital revolution upon the world's technological landscape further in Chapter X. For now, having clarified the meaning of technology and its relation to science, which we'll depend upon throughout, we're in position to begin exploring the historical origins of technology and its co-development with early humans.

Sources / Further Reading

Ferré, Frederick. "Defining Technology," Chapter Two of *Philosophy of Technology.* University of Georgia Press, 1995.

Ihde, Don. "Technology," Chapter Two of *Philosophy of Technology: An Introduction.* Paragon House, 1998.

Mitchum, Carl. "Engineering Philosophy of Technology" (Chapter One) and "Humanities Philosophy of Technology" (Chapter Two) of *Thinking Through Technology: The Path between Engineering and Philosophy*. University of Chicago Press, 1994.

Pitt, Joseph C. "Looking for Definition: Epistemology and Technology," Chapter One of *Thinking About Technology: Foundations of the Philosophy of Technology*. Seven Bridges Press, 2000.

Stanley, Autumn. "Redefining Technology" and "Redefining Significance," from the "Introduction" to *Mothers and Daughters of Invention: Notes for a Revised History of Technology*. Rutgers University Press, 1995.

Volti, Rudi. "The Nature of Technology," Chapter One of *Society and Technological Change*, 8th ed. Worth Publishers, 2017.

Part Two: Technology and Human Development

Chapter III:
Technology and Human Beginnings

HUMAN INVOLVEMENT WITH TECHNOLOGY has an exceedingly long history. In comparison, written history is a recent development. Indeed, the real question facing anyone endeavoring to place a beginning date on human interaction with technology is not, "When did our human ancestors first avail themselves of technology?" but "At what point did the tool-making ancestors of modern human beings assume a distinctly human identity?" There is compelling evidence that **stone tools** (sharp-edged, hand-held instruments for cutting, chopping, and scraping) were being fashioned as early as 2 to 2.5 million years ago. Apparently, they were made by the "percussion method": striking one rock with or against another, breaking off pieces and leaving a sharp, jagged edge. The sharp-edged flakes thus removed probably were put to use also. Anthropologists refer to this preceding as the "**Oldowan tool tradition**" or the "Oldowan tool industry."

The identity of the tool maker is more difficult to pin down. The answer to this question, and to that of similar ones, is subject to debate and revision. The creature that fashioned these tools is believed to be *Homo habilis* and lived between 2.4 and 1.6 million years ago. The cranial capacity of this early member of the genus *Homo* was significantly greater than that of the other members of the biological family to which it belonged, the *Hominidae* (or great apes) which includes orangutans, chimpanzees, and gorillas. *Homo habilis* is also distinguished from the others by the inclusion of meat in its diet; this was probably obtained by scavenging rather than by hunting. Since its teeth and jaws were adapted to a vegetarian diet (as were those of the other members of its biological family), *Homo habilis* could not have availed itself of this important source of protein without sharp cutting instruments. This added dietary resource would have meant greater survivability and greater biological success for the species. The increased intelligence (presumably resulting from the larger brain as well as from better organization of the nervous system), the advantages of tool-making (presumably made possible by the increased intelligence), and the more nutritious diet would have been mutually reinforcing. Greater intelligence also would have fostered more effective planning and better cooperation within groups. Even so, as *Homo habilis* was not a human being, human involvement with technology did not yet exist.

And what of technology? Was the Oldowan tool industry the "first technology," as it has sometimes been called? The tool-making and tool-using activities just described clearly satisfy the first four conditions (set out in Chapter II) for something to qualify as a technology. But do they satisfy the fifth condition? There are good reasons to doubt that these activities were instinctive, so there must have been an initial innovative activity or process of some sort. This leaves only the question of whether there is any evidence of the exploitation of or even the recognition of the potential for further innovation that this tool making and using embodies. Prior to approximately 1.4 million years ago, tools and techniques appear to have been comparatively simple and impressively uniform over vast stretches of time and space. Innovation clearly was lacking during this very early period of tool-making. Technology, that powerful engine of change, evidently had not yet put in an appearance either.

This chapter surveys the most significant periods of tool making and using, along with the more salient features of the human ancestor associated with each of them. The chapter culminates in an account of that burst of technological innovation that occurred in Europe and Africa beginning about 35,000 years ago—accompanying the arrival there of anatomically modern human beings (people biologically indistinguishable from members of the current population). Attention also is given to the appearance of anatomically modern human beings in Africa even earlier, about 100,000 years ago, as well as to some of the evidence for technological innovation that occurred there subsequently. Additionally, we discuss the question of what happened during this period in southern and eastern Asia.

A. More Advanced Tools and Their Maker

Beginning sometime between 1.4 and 1.7 million years ago, much more elegant stone tools begin to show up in the archaeological record. Collectively, these are known as the "**Acheulean tool industry.**" Especially notable is the hand-ax, which has the approximate form of a pear-shaped diamond or an almond with sharp edges around the periphery and a point at the narrow end (see Figure 3-1 below). These hand-axes were made of flint, solidified lava, quartzite, chert, or obsidian—materials that lend themselves to the flaking process. The small flaking scars on these stone tools and their relatively clean lines testify to the skill involved in their production and to the intelligent choice of raw materials.

A fundamental characteristic of the Acheulean tool-making tradition is its multistage nature, including especially the preparation of pieces of stone or "cores" in such a way as to leave a flat area that is roughly perpendicular to an

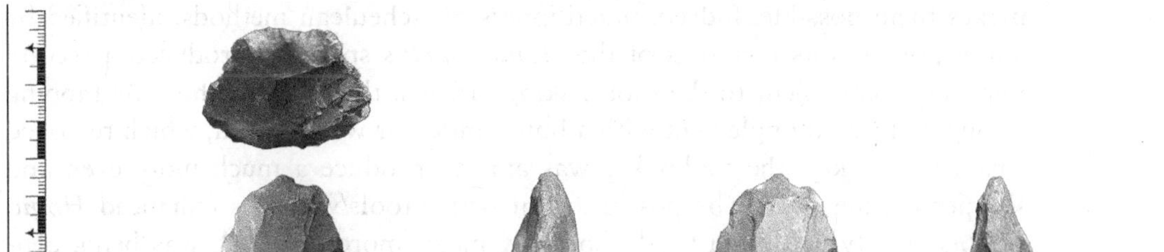

Figure 3-1. Acheulean Hand-ax. The Portable Antiquities Scheme, Matthew Fittock. CC-BY-2.0.

adjoining surface. A well-placed blow to such a "striking platform" near its intersection with the other surface and along a line roughly parallel to that surface will produce a fracture, dislodging a long, thin flake stretching across the surface of the stone being worked. In the Acheulean tradition, this process was applied repeatedly to opposite sides of the stone. The result was a relatively thin, bifacial artifact (having the almond shape mentioned above) with comparatively straight edges, which made it much more effective as a cutting instrument. It should be noted that current understanding of Acheulean tool-making and use (as well as that of earlier ones) has been achieved in part by attempting to replicate them and not merely by studying ancient artifacts.

The significance of the technological development under discussion should not be lost on anyone who has been fascinated by an arrowhead they came across in a cultivated field or other ground or who has marveled at the elegance of the Folsom points. Of course, many methodological improvements and an immense period of time separate those artifacts from this first use of a technique that makes them possible. Indeed, practitioners of Acheulean methods, identified by anthropologists as members of the *Homo erectus* species, introduced a second important refinement to the tool-making process: the **baton method.** By tapping a roughly formed implement with a bone, antler, or wood baton, which removed small, thin flakes, the toolmaker was able to produce a much more even and sharper cutting edge. The possession of better tools/weapons enhanced *Homo erectus*' ability to obtain food, especially meat (more of which was being consumed) and to fend off predators. (The hand-ax is, quite literally, a short dagger.) Members of the *Homo erectus* species were bigger and had a larger cranial capacity than *Homo habilis*, and they are believed to have supplemented the plant food in their diet with meat from hunting rather than just from scavenging.

The extent to which members of the *Homo erectus* species engaged in hunting remains a matter of debate among anthropologists, as does the question of whether they were able to control fire. The latter topic is especially significant because starting and maintaining fires is the first instance of human beings' (or their forebears') gaining control of a natural process. Furthermore, since being able to manage fire makes cooking possible, questions concerning the biological as well as the cultural effects of having cooked food to eat quickly come to the fore. To what extent did eating cooked food contribute to "face reduction" and related changes (the decrease in the length and thickness of the jaws, the size of the teeth, and massiveness of the supporting head and neck structures), a process which shows itself in differences between *Homo habilis* and *Homo erectus*, between early and late *Homo erectus*, and between the latter and *Homo sapiens* (including the Neanderthals)? And what effect did eating cooked food have upon the digestive tract and supporting structures? Such questions are not easy to answer, but some things are known. Cooking food makes the digestion-absorption process more efficient and thus reduces the overall amount of food that needs to be consumed. Accordingly, it is no surprise that the digestive tract of modern human beings is significantly smaller (relative to overall body mass) than that of the great apes or (almost certainly) that of the other (now extinct) members of the genus *Homo*, including those with significant amounts of meat in their diet.

Another important practice that contributed to the biological success of *Homo erectus* (and that may have benefited *Homo habilis* as well) is food sharing: members of a band coming together to share the food they had obtained by gathering, scavenging or (in the case of *Homo erectus*) hunting. Especially significant is the sharing of vegetable foods gathered by females. While there is some very limited sharing of meat among chimpanzees, this sharing of plant foods is unprecedented among primates. Ordinarily they are simply eaten on the spot. Furthermore, the transportation of significant quantities of plant foods would require containers or carrying devices, which are an important but often overlooked part of the human technological repertoire. Since these most probably were made of perishable materials unlikely to have survived from prehistoric times, it is hardly surprising that they—and the crucial role of female gathering—did not begin to receive the attention they deserve until recently.

Perhaps it should be pointed out that intelligence offers significant advantages for gatherers of vegetable foods that are high in nutritional value: the ability to recognize a variety of suitable plants (including those with edible roots or bulbs) as well as poisonous ones, the ability to identify the locales favored by desirable plants and to learn their seasonal rhythms, the ability to develop methods of food preparation, and so on. The greater ability of *Homo erectus* to survive and multiply (as compared to other members of the *Homo* genus) is demonstrated by the species' having established itself beyond the confines of Africa: in Europe, Southeast Asia, and China. Despite the biological success of the species and despite evident refinements in the stone tools produced as time went on, one could not really characterize *Homo erectus* as innovative. The basic process used in making tools remained the same from some 1.4 million years ago until about 200,000 years ago.

B. Yet More Sophisticated Techniques and Survival Strategies

At the end of the period just discussed, a major change in tool-making technology was getting underway: a shift from core technology to **flake technology**, from shaping a given hunk of stone or "core" into a useful implement to preparing a core in such a way that a flake struck from it is a fully formed instrument, requiring only a little touching up. Up to a point, of course, nothing more than a matter of emphasis was involved. Both cores and flakes could be and were put to use. But in the Levalloisian technique, which came into use about 200,000 years ago, the core ceased to be anything other than a left-over hunk of rock. The superior qualities of the flake tools produced and the efficiency of the method account for its success. Its practitioners first worked a flint core until its lateral surfaces resembled elongated tortoise shells then formed a striking platform at one end. Following that, three or four fully formed flake tools were struck from the core. The development of the Levalloisian technique evinces unprecedented mental powers on the part of its practitioners. It is one thing to reshape a piece of rock by chipping away at it bit by bit until it assumes an acceptable form—however, the process is guided: by a physical model, by a mental image (perhaps a memory), by an intuitive sense of "rightness" in one respect or another. It is another thing altogether to have the imagination and forethought to work a core into the form—not of a tool but of an object that has the remarkable property that if struck in just the right way, it will yield flakes that are fully formed tools!

While the exact identity of the creature who employed the Levalloisian technique remains obscure (was it a late form of *Homo erectus* or an early form of *Homo sapiens*, "archaic *Homo sapiens*"?), the practitioners of the next distinct type of stone-tool technology was none other than the Neanderthals (or people closely related to them). By about 100,000 years ago, these early people had developed a more advanced type of flake technology, the **Mousterian** (see Figure 3-2 below). As a first step, they prepared small disk-like cores; next, numerous small flakes were struck from these by repeated blows along their outer rims, directed toward their centers. This new method had three distinct advantages over older techniques: it produced much more cutting edge from a given amount of flint; it enabled knappers (those shaping stone tools by striking the cores with other stones or objects) to produce large numbers of usable flakes with much less effort; and it provided the basis for a greater diversity of flake tools. Practitioners of this more sophisticated flake technology became quite adept at retouching and sharpening flakes into an impressive variety of tools for work on wood, carcasses, and hides. Because of the many types of points suitable for hafting (mounting on

Figure 3-2. Mousterian Cutting Tool. Image by Guérin Nicolas. CC-BY-SA 3.0

a wooden shaft) found at Mousterian sites, it is likely that their makers employed thrusting spears.

Living in Europe during one of the cold periods (the last ice age which set in about 70,000 years ago), the Neanderthals of necessity relied upon meat as their primary source of nourishment, vegetable foods being available only seasonally or sporadically. And there is solid evidence that they preyed upon large animals, such as horses and bison, most notably by herding them over cliffs or into ravines, a tactic also employed somewhat later in North America by the Native Americans. Their utilization of this method is a testament to their communication and organizational capabilities. An important benefit of hunting, especially of big game, is the animal hides thus obtained. Once they learn to clean and process them properly, people can use them to shield themselves from the cold, whether by draping them over their bodies or using them in the construction of shelters. While there is as yet no direct evidence that the Neanderthals availed themselves of such possibilities, they did leave behind stone tools well suited for scraping and piercing, the very operations one would expect people to engage in if they were fashioning useful items from animal hides. In any case, there is solid evidence that the Neanderthals practiced the art of fire management, which would have been an enormous advantage to them, not only for keeping warm but also for keeping large, dangerous animals at bay, especially the former occupants of caves in which the Neanderthals had sought shelter. Additionally, as noted above, being able to maintain fires makes cooking possible. The Neanderthals and other populations of archaic *Homo sapiens* also used fire to make their wooden spears more effective by hardening them (heating dehydrates the wood).

The success of the Neanderthals and other populations of archaic *Homo sapiens* in challenging new environments demonstrates that they were capable of developing the requisite new technical and social skills. Furthermore, late in the period during which they were present in Europe (which ended about 33,000 years ago), the Neanderthals took up the more sophisticated blade technology (discussed below) practiced by the anatomically modern human beings then moving into the continent. Even so, the status of the Neanderthals remains controversial. Disagreement persists regarding such questions as how they are related to people living in Europe today and whether they truly were human. In any case, they really were not a very innovative people. On the contrary, the most salient feature of their technology and, insofar as one can tell, their culture is sameness—across vast spaces and temporal periods. The stone tools they left in Spain differ little from those left in Uzbekistan near the Afghanistan border, most of those from 35,000 years ago differing little from those from 100,000 years ago. Admittedly (as noted), things may have moved a little faster toward the end of this vast period; but following it, implements and activities became progressively more complex and innovation incomparably more rapid—so much so that this transition has been referred to as "the Big Bang" and "the Great Leap Forward." It marks the beginning of what anthropologists call the "Upper Paleolithic" period, the last segment of the Old Stone Age. Most importantly for present purposes, not only was technological innovation undeniably present during his period but much (if not most) of it involved the realization of possibilities that would not have existed in the absence of older technologies.

C. Technology and Human Beings—At Last!

This great watershed in human affairs coincided with the arrival in Europe about 35,000 years ago of anatomically modern human beings, people biologically indistinguishable from members of the current population. Based on the name of the place where their remains were first discovered, they are often referred to as the "Cro-Magnons." That something very special was underway is attested by the multi-dimensional character of developments. Not only did people begin to practice powerful new methods of making stone implements, to employ a greater variety of and much more specialized tools, and to extensively utilize materials (bone, antler, ivory, etc.) previously of only marginal importance, they also began, with dramatic suddenness, to carve exquisite ivory figures and to paint, on cave walls, pictures of animals so stunningly executed that, upon their rediscovery in the late nineteenth century, they were dismissed initially as modern forgeries! This new capacity for aesthetic expression also seems to have manifested itself in music making: by no later than 25,000 years ago, "penny whistles" with finger holes were being fashioned from hollow bones. While there is evidence that the predecessors of modern human beings had a capacity for aesthetic appreciation (of flowers, seashells, mineral crystals, etc.), there is little to suggest a capacity for aesthetic expression. In this regard, one can do little more than point to the symmetry and graceful lines of Neanderthal implements and to their use of such natural pigments as red ocher and black manganese, which may or may not have involved aesthetic expression.

Care should be taken at this juncture not to assume that the transition under discussion came about simply as the result of anatomically modern people being able to do things that their predecessors simply could not do. As will be discussed in Chapter VIII, social and environmental factors can exert a powerful influence upon whether there is technological innovation or stagnation. Evidence that more than a simple heritable factor is involved in this transition is afforded by the example of one of the hallmarks of Cro-Magnon culture: blade technology. The "blade technique" is a method whereby long, thin, blade-shaped flakes are repeatedly struck (each by a single blow) from a specially selected and prepared (roughly) cylindrical core. (Flakes are considered to be blades if they are at least twice as long as they are wide and have their long edges roughly parallel.) The blade technique has two distinct advantages over earlier methods: efficiency and speed. Many more blades than Mousterian flakes can be obtained from a given amount of flint, and a blade's cutting edge is four or five times more usable than a flake. Furthermore, a skilled stoneworker can convert an entire core (or nearly all of it) into a pile of blades in a matter of minutes!

In Mexico, for example, where the blade technique survived into historical times, an expert Aztec knapper was observed producing some one hundred usable blades in less than an hour. The method by which this remarkable result is achieved is itself a wonder. The stoneworker begins by forming a roughly cylindrical core with a small ridge extending down one side from one end to the other. Then by directing a blow at the edge of one end of the cylinder in line with the ridge, he is able to detach from the core a long, thin sliver with razor-sharp edges. The removal of a blade from a core in this manner leaves a shallow, rounded groove in place of the ridge. The procedure can then be repeated, taking advantage of the ridges on each side of the groove, and repeated again and yet again, progressing around the core, which assumes a fluted appearance. This process, which has been compared to removing leaves from an artichoke, can be continued until very little of the core remains.

The blade technique is not a simple skill. Proficiency in it can be attained only through an extended learning process and much practice and can be maintained only by continued practice, as is the case with any high skill. Of course, the existence of such high-level skills from the time of the great leap forward (roughly 50,000 years ago, when we see evidence of rapid advancement towards more complex modern human-like behavior) is compelling testimony of the capabilities of the people involved (see the Levalloisian technique above). The Neanderthals' having begun to practice the blade technique following the arrival in Europe of anatomically modern people (as noted above) testifies to their capabilities and weighs heavily against any claim that this innovation was the result of specific genetic change(s) directly related to it. Furthermore, stone blades (evidently produced in the manner described above) that are about 240,000 years old have been discovered in Kenya. Since anatomically modern human beings are believed to have first appeared in Africa about 100,000 years ago (and surely no earlier than 200,000 years ago), it follows that these blades must have been made by archaic *Homo sapiens*, people rather like the Neanderthals. However, these earlier people did not make blades the foundation and centerpiece of their tool making as did the Cro-Magnons, who fashioned the blades into an impressive variety of specialized tools, as is discussed below.

A second basic element of Upper Paleolithic technology is **pressure flaking.** Instead of striking the piece of stone being worked, either directly or by means of a punch, practitioners of this technique induce fracturing by pressing a sharply pointed piece of antler, bone, or wood against it. This concentrated pressure causes small, thin flakes to split off. And therein lies the great advantage of pressure flaking. Because the chips removed in this manner are small and thin, employing this technique, especially in the finishing stages, enables stoneworkers to exercise more precise control over the flaking process and to produce more delicately shaped implements. The celebrated willow-leaf and laurel-leaf blades and points (the longest recovered is thirteen inches long but only a quarter of an inch thick!) were fashioned in this manner. Only a few naturally occurring materials—obsidian (volcanic glass), jasper, and a good grade of flint—lend themselves to pressure flaking. Since these materials are by no means readily available everywhere, the technique could not have been exploited by Upper Paleolithic people to the extent that it was unless they were capable of carrying on trade with the inhabitants of neighboring territories or of mounting expeditions to distant sources of special materials not available locally. Flint blade tools have been discovered at a major archaeological site on the Don River in Russia that is at least 90 miles from the source of the flint. The use of materials transported from distant sources and the more complex social organization necessary for such undertakings is another important element of the Upper Paleolithic period. Earlier peoples, including the Neanderthals, relied on materials that could be obtained locally; they did not engage in long-distance trade.

Having an efficient means of producing large numbers of high-quality stone blades and an effective way of shaping them with great precision, the Cro-Magnon people of the Upper Paleolithic period were in a position to produce and make use of a great variety of specialized stone tools—which is just what they did. It appears that the Cro-Magnon people made and used over a hundred distinct types of tools. This total may not seem significantly larger than the sixty or so types of tools that archeologists have distinguished in the Neanderthal toolkit: scrapers (presumably intended to be held horizontally), double-edged knives, knives with one edge (the "back") blunted (so finger pressure can be applied), notched or toothed flakes (possibly for shaping wood), etc. Such a comparison may be misleading, however. While an individual Neanderthal stone implement may appear (to modern eyes) to be a distinct type of tool (a scraper, a knife, a hand ax), Neanderthal implements as a whole vary almost continuously in shape and size. Furthermore, microscopic studies of the wear marks on Neanderthal stone tools reveal that they were used to cut animal hides, meat, and bone as well as wood and the non-woody parts of plants, and that a given stone implement (no matter what its size and shape) was likely to have been used to cut any or all of these things. Consequently, the extent to which the Neanderthals availed themselves of the advantages of specialized tools remains unclear.

In contrast, it is well established that the Cro-Magnon people used a large number of quite specialized tools. While it is neither necessary nor desirable for present purposes to undertake a systematic review of these tools, it is important to examine those that exemplify the realization of potentialities inherent in pre-existing technologies, a crucial element in the burst of technological innovation that characterized the Upper Paleolithic period and in subsequent episodes that have repeatedly remade the human world.

One specialized tool that played a central role in the Upper Paleolithic burst of technological innovation is the **burin**. Also important is the perforator. The burin is a chisel-like tool with a strong, sharp corner or point that can be used to work hard materials. One can make deep scratches in their surfaces by applying pressure as the instrument is drawn across them. Thus, the burins of the Upper Paleolithic period are similar in function to those of modern engravers of metal and stone. By making a deep scratch around an antler, bone, or piece of ivory and then subjecting it to pressure or a well-placed, sharp blow, one can produce a clean break at that location. This same technique is still used by glaziers who "cut" glass by making a scratch on its surface and snapping it in two along the scratch. By repeatedly drawing a burin along the same line, one can make a deep cut in a piece of material—even cut completely through it. By making two such parallel cuts in a length of hollow bone, one can remove a piece suitable for shaping into a needle. The Cro-Magnon people were quite expert at using the burin to work bone, antler, and ivory into a variety of useful and elegant implements, including needles, (conical) spear points, harpoon heads (some with a single row of barbs, others with two), and elaborately carved atlatls or spear throwers (devices that, in effect, increase the length of the arms of those hurling the spears and thus the velocity and range of the missile). The perforator is a sharp-pointed instrument that can be used as an awl, to punch holes in leather for example. The Cro-Magnons also used them to make holes in bone and ivory in the production of beads and needles. Typically, holes were "dug out" on the opposite sides of a piece of material until the holes met. (Properly speaking, perforators were not really drills. Those came later.)

Surprisingly, the Cro-Magnon people were not the first to employ the burin; it is found in some Neanderthal and even earlier tool assemblages. Furthermore, as noted above, the Neanderthals left some tools that appear to be suitable for piercing animal hides, and they made some use of antler and bone tools, e.g., in the baton method described earlier in section A. But in comparison to what came later, the Neanderthals' utilization of the burin, tools suitable for making holes, and materials such as bone and antler was extremely limited. About one in forty of the tools brought to light in the excavation of a typical Neanderthal site are bone; the remainder are stone. At a Cro-Magnon site there may be as many of the one as of the other. Somehow, the Cro-Magnon people succeeded (and the Neanderthals did not) in realizing the possibilities presented by the burin and perforator in combination with materials like bone, antler, and ivory. Similarly, while the Cro-Magnons made extensive use of compound tools, the evidence for the use of such devices by the Neanderthals is confined to stone spear points suited for hafting on wooden shafts. The significance of this difference will not be lost on anyone who has tried to cut even a small branch with an ax head rather than an ax.

An important feature of the Upper Paleolithic period emerging repeatedly when powerful new technologies burst upon the scene is a marked increase both in the overall human population and in the size of the communal groups in which people live. As the period unfolded, human populations increased enormously—by as much as tenfold in some parts of the world. The Cro-Magnon people were able to sustain these greater populations because of their more effective hunting methods and weapons. (And where conditions were favorable, Upper Paleolithic people were quite proficient at "harvesting" wild grain and other vegetable foods.) Analyses of animal remains at their numerous living sites reveal that they were much more successful as hunters than their predecessors, not only in systematically preying upon large mammals but in exploiting food sources not previously tapped: fish, birds, and small mammals too elusive to be taken on a regular basis by older methods. Upper Paleolithic people also lived in groups that were two to three times as large as earlier peoples. (Neanderthal bands are believed to have averaged about twenty-five members.) These larger communal units reflect both greater exploitation of food sources within range of such a group and the requirements of their new hunting practices, especially the pursuit of large herd animals.

D. Sewing, Spinning, and Fishing

Not only did the powerful new technology enable Upper Paleolithic people to tap new resources where they already lived, but also made it possible for them to move into vast areas previously uninhabited. Having invented the needle and sewing, people were able to make snug, form-fitting Inuit-type clothing. Thus protected, they were able to survive the brutal winters of northern Russia and Siberia. From there, they migrated into the Americas by way of the Beringia land bridge, which lay exposed by the lowered ocean levels of that glacial period. Sewing, of course, requires not only needles but also thread or something rather like it. But since such materials are highly perishable, knowledge of their origin and development is extremely elusive. Those parts of animal carcasses, such as tendons and intestines, which lend themselves to such purposes were surely put to use whenever they were available and when cords and strings were needed. Even today, some musicians still prefer catgut for their stringed instruments; workers process and stretch the intestines of (primarily) sheep to produce it. For certain purposes, such as snares (traps of a certain type), strips of animal hide ("rawhide") were used.

As is often the case with useful materials or devices, the full potential of cord or string-like things was not realized until a way was found to increase the supply significantly. Sometime between twenty and thirty thousand years ago, people (probably women, given such factors as the earliest uses of the product; see below) learned to make long, tough threads by twisting wads of tiny, weak fibers together. They invented spinning! Miraculously, a tiny sample of threads that had been spun from plant fibers and then twisted into a cord some 15,000 years ago has been found in the caves at Lascaux, France.

Evidence from earlier times is more indirect, however. For example, during the Gravettian period (from 26,000 to 20,000 years ago), needles came into common use (implying the greater availability of thread), and the eyes of needles and the holes in beads became smaller (suggesting greater reliance upon thread as compared to sinew or gut as well as more highly developed thread-making skills, enabling people to produce thread that was thinner but also strong and even). Neat rows of small beads stretching across the skeletal remains of individuals buried during this period must surely have been sewn on their clothing, which was probably fashioned from animal skins that disintegrated long ago. Further evidence that thread was being made by twisting tiny fibers together is afforded by the detailed engraving of a garment worn by a bone figurine that clearly shows not only the twists in each of the strings comprising it but also the individual fibers in their frayed ends. This tiny "Venus" figure was found at Lespugue, France, and is believed to have been carved about 20,000 years ago. The Venus figurines of this period have exaggerated breasts, buttocks, and bellies and more often than not are completely naked. The garment portrayed is quite small: a band around the hips from which long strings hang, but only from the central portion in the back. A similar figurine from Russia of roughly the same age sports a similar garment but with strings that are shorter and hang only in the central portion of the front. Since these "string skirts," as they have been called, served no practical purpose (they offered neither warmth nor protection of the wearer's modesty), they must have had another function. For obvious reasons, it is thought that they were associated with women's child-bearing powers.

In addition to sewing, string or cord has numerous practical applications. Among the more important are **nets** for catching both fish and birds and **lines** for catching fish with bait. The result of these inventions was the addition or major expansion of important sources of food. Indeed, the presence of many more fish and bird bones in the upper levels of refuse piles than in the layers below is good evidence that people have learned to make string and cords from plant or other natural fibers. And, perhaps most profoundly, once thread becomes available, weaving cloth becomes possible. But that possibility is not one of the potentialities of thread that was realized by the Cro-Magnon people, nor could they have imagined the opportunities for aesthetic expression that weaving would offer.

E. Aesthetic Expression, Symbolic Representation, and Bodily Ornamentation

But the Cro-Magnons most certainly did take advantage of the opportunities for aesthetic expression afforded them by the technological innovations they did make—and with marvelous results! The elaborately carved atlatls and rows of beads sewn on garments have been mentioned already. An example that is perhaps even more impressive is the exquisitely carved ivory Vogelherd horse (Figure 3-3 below). What is truly remarkable about this tiny figurine is that it is both (i) as old (some 30,000 years) or older than any other human rendering of nature and (ii) clearly the work of an accomplished master. Surprisingly, there is no evidence of a preceding tradition producing cruder, more schematic figures, as one might expect. Clearly, this Upper Paleolithic artisan had remarkable powers of perception, expression, and execution that any artist today might covet. Another remarkable feature of the tiny equine figure is the clear evidence of much handling. A microscopic examination reveals that its finely-carved mouth, nose, ear, and eye have been worn down, as occurs with much-circulated coins. This evidence of ample handling suggests that it probably was used in rituals or ceremonies.

Another venue for aesthetic expression during the Upper Paleolithic period and beyond is the human body itself. There is abundant evidence of bodily

Figure 3-3. The Vogelherd Horse. Carved from mammoth ivory over 30,000 years ago. Image credit to Mogadir. CC-BY-SA 3.0.

ornamentation from the early Aurignacian period (34,000 to 30,000 years ago): objects that appear to be beads or pendants made from soft stone, shells, animal teeth, and (most of all) ivory from the tusks of mammoths, many of them perforated for suspension, have been found at hundreds of sites. While it is not known precisely how the beads and pendant-like ornaments were employed during this early period (were they hung about the neck, sewn onto clothing, or used in some other way?), excavations of a burial site in Russia, dating between 28,000 and 24,000 years ago, have revealed thousands of ivory beads and pierced animal teeth arranged in rows as if they had been strung. In contrast, even though the Neanderthals may have painted their bodies (they seem to have employed pigments in powder form and in stick-like lumps, some of them bearing signs of having been rubbed on a soft material like human skin or leather), they do not appear to have worn beads or to have adorned themselves with animal teeth or claws, as have more recent peoples. (No beads or perforated animal teeth have been recovered from any Neanderthal site.)

The renderings of animals in ivory and on cave walls are not merely the first clear examples of aesthetic expression: they are also the earliest unequivocal instances of human symbolic activity. While the Neanderthals engaged in activities that can be interpreted as having a symbolic character, the evidence is too sketchy to reveal anything about the specific nature of the symbolism. Thus, the extent and character of Neanderthal symbolic activity remains essentially unknown. However, there can be no doubt that the Vogelherd figurine is a symbol of a horse or that the lifelike paintings of horses and bison in caves represent those animals. And there are good reasons to think that these images had important places in ceremonies or rituals of some kind even if we may never know very much about them. (In addition to the wear on the horse figurine, this evidence includes clear signs that depictions of animals in caves have been repeatedly retraced.) There are also reasons to believe that the beads and pendants described above served some purpose in addition to decoration. Evidently, the people of the early Aurignacian period attached great importance to these ornaments: not only was making the ivory beads, for example, a time-consuming, multi-step process, but these people seem to have transported certain types of soft stone, shell, and ivory to processing sites hundreds of miles from the places where they were naturally available. Consequently, it is highly likely that these ornaments served some such purpose as social display or badge of status, as would be expected in an increasingly complex society capable of carrying on trade with the inhabitants of neighboring territories or mounting expeditions to distant sources of special materials not available locally.

F. Cultural and Geographical Variation

There is one more feature of the technology of the Upper Paleolithic period that should be noted because it is a harbinger of a major feature of the human scene thereafter. This characteristic has its source in the capacity for innovation which human beings somehow had acquired by that time. Given that capacity, it is inevitable that groups living in geographical separation will develop distinct ways of doing things. The result, of course, is cultural variation, which has been an important feature of human societies ever since. It showed itself right away in the tool-making traditions of the Cro-Magnon peoples. In France alone, no less than five major such regional traditions flourished during the 25,000-year period ending about 10,000 years ago. (And some experts distinguish several divisions within these.) Furthermore, nothing inherent in the raw materials, the production process, or the function of these implements necessitates their having the characteristic features that identify them with a specific tradition. Rather, there is every reason to believe that tools, having one or another of these sets of features, were made that way over an extended period (some thousands of years) because generation after generation of their creators were taught to do so. Thus, these early styles of making tools are among the world's earliest cultural traditions.

This discussion has focused upon developments in Europe during the Upper Paleolithic period because that is where much more material seems to have been preserved and where much more study has been done. However, neither the burst of technological innovation that took place during this period nor its effects was confined to Europe. There are good reasons to believe that a transition took place in Africa paralleling the one that took place in Europe, including compelling evidence of a marked enhancement in the hunting capabilities of the anatomically modern people living in Africa between their first appearance there roughly 100,000 years ago and about 18,000 years ago. This conclusion was reached on the basis of studies of animal remains recovered from sites occupied by human beings during this period. Excavations in caves along the southern coast of Africa indicate that the diet of the earliest modern people living there included seals, penguins, shellfish (like the limpet), antelope-sized mammals, and young or aged buffalo. Notable in their absence are significant numbers of fish, flying birds, pigs, adult buffalo (other than the aged), and other large mammals. In contrast, by about 18,000 years ago the inhabitants of these caves were successfully preying upon now-extinct giant buffalo and hartebeest; the former weighed some one-and-three-fourths tons, the latter about one ton. They also hunted bush pigs and wart hogs. The latter are aggressive animals with fearsome tusks. They travel in packs and will turn upon would-be predators *en masse*. The remains of fish at the site testify that the art of fishing had been mastered by this time also. As people elsewhere were doing, they probably used a fish gorge, a short length of bone or wood attached at its middle to the fishing line that baited like a fish hook. Once swallowed, a tug on the line turns it crossways in the fish's throat so that it can be pulled in. Lastly, it should be noted that it remains an open question whether there ever was a watershed in Southeast Asia or East Asia similar to that being discussed. Developments there may have followed a very different pattern.

* * *

The nagging question lurking in the background of the last few pages is, of course, "Exactly what happened around 35,000 years ago that did not happen 100,000 years ago when anatomically modern people first appeared?" In short, what caused the emergence of the tool-making cultures discussed above? One suggestion is that complex languages of the kind that now exist first appeared then, and that these made it possible for people to think about problems, objects, and processes in ways they previously could not. Having a word for a property or thing makes it possible to think about it and its possible effects when it is not present. Furthermore, having words for properties and things makes it possible to think about novel combinations of them, a capability clearly vital to technological innovation. But as these two operations also can be carried out by visualization, the question becomes a matter of whether the addition of language could have enhanced such human capabilities sufficiently to account for the differences under discussion—a very hard question to answer.

In any case, since spoken language leaves no physical residue, it is difficult, if not impossible, to obtain hard evidence of such a transition. Given the relative absence of innovation on the part of the Neanderthals, it is hardly surprising in the present context that attention would be focused on the question of whether the Neanderthals used fully formed language(s). There has been much discussion and debate concerning whether they were physically capable of producing the full complement of sounds employed in languages as we know them, but much uncertainty remains. At this point, it does not appear that this topic is likely to provide a basis for significant insights related to the fundamental question raised just above.

While it is tempting to offer a linguistic explanation for the burst of innovation described above, it is worth spelling out further just what the alternative might be. Visualization, mentioned above, could serve as the basis for imagining new uses of tools and modifications that might be made, but recent research also suggests that certain ways of structuring learning environments could also provide conditions ripe for innovation even in the absence of language. For example, Kim Sterelny argues that early humans' evolved capacities for trial-and-error learning, in the right social environment in which children and young adults are permitted to observe and attempt tool creation and use in competent adults, can lead to information transmission between generations without much in the way of explicit instruction. Feedback loops may then emerge between individual cognitive capacities (learning ability and capacities such as imagination) and social organization (such as the way the young are allowed to participate in tool creation) under the demands of changing and diverse environments that help drive the development of humans' unique capacities, including those cognitive abilities that result in technological innovation.

Whatever turns out to be the explanation for the fundamental changes that took place around 35,000 years ago, one thing is clear: the capacity for technological innovation is closely tied to the other elements of the human equation: aesthetic expression, the use of symbols, bodily adornment, complex social and economic organization, and cultural variation. Fully developed language is also part of that equation, the only question being whether its advent preceded or, like the others mentioned, was part of that long-ago starburst.

Sources / Further Reading

Barber, Elizabeth Wayland. *Women's Work: The First 20,000 Years.* W.W. Norton, 1994.

Campbell, Bernard G., et al. *Humankind Emerging*, 9th ed. Pearson, 2005.

Diamond, Jared. "The Great Leap Forward." *Discover*, vol. 10, no. 5, May 1989, pp. 50–60.

——. *Guns, Germs and Steel.* W.W. Norton & Company, 1997.

Haviland, William A. *Anthropology*, 8th ed. Harcourt Brace College Publishers, 1997.

Johanson, Donald, and Blake Edgar. *From Lucy to Language.* Simon & Schuster, 1996.

Leakey, Richard, and Roger Lewin. *Origins Reconsidered.* Doubleday, 1992.

Marshack, Alexander. "Exploring the Mind of Ice Age Man." *National Geographic*, vol. 147, no.1, Jan. 1975, pp. 64–89.

Putman, John J. "The Search for Modern Humans." *National Geographic*, vol. 174, no. 4, Oct. 1988, pp. 438–77.

Schick, Kathy D., and Nicholas Toth. *Making Silent Stones Speak: Human Evolution and the Dawn of Technology.* Simon & Schuster, 1993.

Sterelny, Kim. *The Evolved Apprentice.* MIT Press, 2012.

Toth, Nicholas. "The First Technology." *Scientific American*, vol. 256, no. 4, Apr. 1987, pp. 112–21.

Walker, Alan, and Pat Shipman. *The Wisdom of the Bones.* Alfred A. Knopf, 1996.

Weaver, Kenneth F. "The Search for Our Ancestors." *National Geographic*, vol. 168, no. 5, Nov. 1985, pp. 560–623.

White, Randall. "Visual Thinking in the Ice Age." *Scientific American*, vol. 261, no. 1, Jul. 1989, pp. 92–99.

Chapter IV:
Woodcraft, Workshops, and Woven Wonders: Technology and the Settled Way of Life

As NOTED IN THE Introduction and Chapter III, social and environmental conditions exert a powerful influence upon whether and to what extent technological innovation takes place. In particular, one of the most important factors that bear upon technological innovation is **sedentism**, having a more or less permanent place of residence. Before they took up the practice of agriculture, people sustained themselves by gathering the fruit, nuts, seeds, roots, etc. of wild plants and by hunting, trapping, foraging, and fishing. Typically, people who meet their needs this way have lived in small, cooperative groups, sharing the food which the members have collected. Such groups remain camped in one spot only so long as local food sources permit. When these begin to thin out, they move on, taking all their possessions with them. Accordingly, such a way of life can hardly fail to be nomadic. Even so, when conditions do permit, they may remain at the same site for extended periods, perhaps an entire season.

In sharp contrast to this usual pattern, a few non-agricultural societies have nevertheless managed to establish themselves in locations where the creatures upon which they depend for food literally come to them! People living along the migration routes of grazing animals, marine mammals, fish, or birds do not have to travel great distances to find prey—at least not during those times of year when migrations are taking place. Furthermore, if they succeed in devising methods of preserving and storing the bounty thus provided them, they can be assured of an adequate food supply throughout the year. Learning to lay up provisions for future consumption constitutes a major departure from the usual ways of foragers like the San of South Africa and the Australian Aborigines, who have extremely limited food-preservation capabilities and usually consume any food brought into camp within a couple of days. Furthermore, taking up the practice of preserving and storing large quantities of food commits people to a **sedentary or semi-sedentary way of life** (having a more or less permanent place of residence), bringing an end to nomadism. Food supplies extensive enough to sustain people for months at a time are too heavy and too bulky to be carried about and are likely to require being guarded and cared for.

Assured of sustenance and freed of the necessity of moving, which not only consumes much time but also limits human possessions to what can be carried from one campsite to another, people living in favored locations of the type described have been able to avail themselves of a great variety of tools, techniques, and materials and to develop very elaborate social structures and cultural traditions, including art forms that would be ornaments to any society. One people thus blessed with the bounty of nature were the Magdalenians, who lived in Europe between 18,000 and 12,000 years ago. Fed by vast herds of large mammals—horses, reindeer, bison, etc.—and by hordes of fish (especially salmon) surging upriver to spawn, these Cro-Magnon people brought Upper Paleolithic European technology and art to its highest level. They added specialized new instruments to the human tool kit (the harpoon and fine bone needles), produced elaborately carved implements from antler, bone, and ivory, and painted the dazzling murals in the caves at Lascaux and Altamira.

Native Americans living along the Northwest Pacific Coast prior to the arrival of the Europeans occupied a similarly favorable environment: a land relatively free of temperature extremes (heat or cold); a land of abundant rainfall, magnificent forests, and mighty rivers; a rugged coastland of inlets, islands, straits, and sounds, ideally suited to a people who had mastered the art of waterborne transport. While nature offered many riches, fishing and sea-hunting were the economic mainstays of the Native American societies of the region. First, there were the hordes of salmon, eulachon (candlefish), and smelt that surged through the straits and up the rivers to spawn and the herring that crowded into the shallow coastal waters to lay their eggs. Since this enormous bounty was available only at certain times of the year, great quantities of these fish were "harvested," dried or smoked, and stored for future use. The second pillar of the Northwest Coast economy was the hunting of marine mammals: sea otters, seals, sea lions, porpoises, and even whales. Exploitation of this resource required not only water craft but also an extensive knowledge of animal migration patterns.

Because of their geographical isolation, the Native Americans of the Northwest Pacific Coast historically had only limited contact with agriculturalists. Furthermore, European sailors did not arrive in the area until late in the eighteenth century, and traditional ways of living persisted in some places at least until the beginning of the twentieth century. Two additional factors enable our incredibly detailed knowledge of the traditional Northwest Coast way of life: firstly, the detailed diaries, memoirs, and drawings of a number of the Europeans who first made contact with the Native people of the region, and secondly, archaeological excavations in the area that have enjoyed unparalleled success. Most notably, excavations at Ozette, Washington led to the recovery of an entire plank house and contents in a marvelous state of preservation. The house is part of a Makah village believed to have been buried in a mudslide some three or more centuries ago. The Makah are Native Americans who have lived in the northwest corner of the Olympic Peninsula for a very long time. One of their most notable traditional activities is whale hunting.

The traditional Northwest Coast way of life offers an unparalleled opportunity to confront in concrete terms the technology, aesthetic expression, social organization, and cultural richness of which human beings are capable when they do not always have to be on the move in search of food but have not taken up the practice of agriculture. It is thus ideally suited to be the first case study of this chapter. We will outline the basic features of the Native Americans' economy and technology, with special attention to the implications of sedentism, and examine their technologies in considerable detail in order (i) to fully appreciate the achievements they represent, (ii) to see clearly how one technology can give rise to another, and (iii) to understand how there can be—latent in the structure of a new technology—new modes of aesthetic expression and forms or processes that become the subject of important scientific investigation. For example, the "twining" method of weaving in the Pacific Northwest derived from basket-making and, like weaving in general, embodies a structure in which geometrical forms are latent and available for aesthetic expression. Section C of this chapter is devoted to Northwest Coast weaving, section B to the truly amazing woodworking done there, and section A to the techniques employed to harvest, process, and preserve the foodstuffs supplied by nature.

A. Gathering Nature's Bounty: Survival and Status in the Pacific Northwest

The Native Americans employed an impressive variety of methods to catch the migrating fish upon which their survival depended: elaborate weirs and fish-traps in the rivers and sometimes along the coast, long stretches of gill nets (with mesh allowing a fish to push its head through just enough for its gill covers to become entangled), trawl lines festooned with baited hooks, huge trawl nets pulled by canoes, seines (fishing nets that hang in the water by using floats on the top and weights on the bottom), dip nets (bags of netting—some quite large—attached to frames with handles), harpoons (spear-like implements with detachable heads to which a line was affixed), leisters (spear-like devices with two springy arms fitted with inward pointing, barb-like spikes), and, of course, the baited fishhook on a line. The great variety of devices employed in the Pacific Northwest constitutes both a testimonial to human inventiveness and a reminder of the consequent non-static character of human existence. Not only is there evidence of the appearance of new technologies in the region's archaeological record, but stories of inventions being made circulated among the Native Americans.

For example, an ingenious bag net used in harvesting eulachon (candlefish) is said to have been invented by a Haisla girl whose inspiration came from watching a bullhead (sculpin) gulping down trout fry. (The Haisla are one of a group of Northwest Coast peoples whose traditional homeland is the northern half of the eastern coast of Vancouver Island and the facing coast of mainland British Columbia extending north about halfway to the Alaska border.) Just as the body of the sculpin tapers back from its large, flat head, so the long, tubular bag net tapers back from its wide mouth. Placed facing downstream in a river where the tide runs swiftly, these nets fill (under favorable conditions) with eulachon (candlefish) surging upriver to spawn. The mouth of the net is then secured and the narrow end hauled aboard a canoe and untied, dumping the catch into the boat. Whether or not this story is literally true, it certainly demonstrates an awareness of the possibility of human beings' inventing things. At the same time, parallels to the Greek myth of Prometheus endowing human beings with fire circulated among the inhabitants of the Northwest Coast. For example, according to one narrative, Raven (the "trickster" character in the lore of the Pacific Northwest) taught the people to build weirs and to construct fish traps.

The wooden fishhooks of the Northwest Coast people testify both to their wood-working expertise and to their spirit-infused way of life. Using steam to soften the wood, the Native craftspeople fashioned fishhooks of bent wood and fitted them with bone barbs. These devices were both functional and elegant, and some were intricately carved. In one preserved specimen, a strategically carved mouth line and an inlaid eye disk bestows upon the hook the overall look of a fish—transforming it into a lure! Small figures of frogs carved on the shank of another hook probably had a similar purpose, but they almost certainly were intended also to tap the powers of the spirit realm. The same is surely true of the small human figures carved on the shanks of some of the fishhooks. Indeed, the entire business of catching and utilizing fish was permeated with spiritualistic elements. The Native Americans were keenly aware that the migrating fish and animals were a special boon, and they naturally assumed, as people typically do in such cases, that they were somehow the intended beneficiaries of this bounty. Since the fish and animals came of their own accord to places where they could be taken easily, the Native Americans assumed that the spirits involved were intentionally providing them with the food supply upon which their very lives depended. Not surprisingly, they were most anxious not to offend these spirits and thus were careful to show them proper respect and to observe elaborate protocols in dealing with them.

The fishing methods used and the processing procedures followed varied, depending upon the location, the kind of fish involved, their life cycle, and the customs of the people. For example, although their favorite quarry was the whale (see below), fish was the mainstay of the Makah diet, as was generally the case in the region. The Makah used the **hook-and-line method** in their fishing activities, which were focused on halibut, a type of flounder. Rocks attached to baited, bent-wood hooks kept them on the bottom, while at the surface, the lines were attached to inflated bladders. After setting several of these devices, the fisherman had simply to watch for the telltale motion of a bladder, signaling that a fish had been hooked, and then hauled in the catch.

Once brought ashore, the fish were cleaned, gutted, and skinned, and their heads and backbones were removed. Typically, this work was done by women. Then the flesh of the fish was cut into thin slices to facilitate drying, which was accomplished by spreading the slices of fish on the relatively flat roofs of Makah dwellings or by hanging them on drying racks (horizontal poles suspended between uprights). The skins were dried also and used to wrap the strips of flesh before they were packed in baskets and stored in the Makahs' houses. These dried strips required no further processing but simply were broken into pieces and dipped in whale oil before being consumed. The utilization of the dwelling as a workshop/warehouse is an important feature of sedentism which persisted until the Industrial Revolution.

Catching and preserving salmon, the most abundant and most economically important fish in the entire region, was approached in a quite different manner.

As the fish made their spawning runs up the rivers and streams, they were harpooned, scooped up in nets, or captured by means of elaborate weirs and fish traps. The salmon were then taken to the Native Americans' homes, where the women cleaned them, removed the entrails, heads, and tails, then split the fish and removed the backbones. If the fish were intended for immediate consumption, they were mounted on roasting sticks set in the ground around a fire and broiled. If the intent was to preserve the fish for future use, they were hung on poles in the houses to smoke-dry. (Since Native American dwellings lacked chimneys, smoke from open fires was allowed to escape through a "smoke hole" in the roof or between the roof planks. This arrangement provided a made-to-order place to cure fish.)

Hordes of herring swarming into shallow coastal waters to spawn afforded another opportunity for Native Americans from the Strait of Juan de Fuca north to obtain food for immediate consumption or to replenish their depleted stocks. Since the herring spawned in the spring when winter stores were likely to be running low, their arrival was particularly welcome—and not just for their flesh but for their eggs, which were especially relished. In the spawning areas the herring were so thick that a special device called a "herring rake" was used to snag several of them at once and dump them into the fisherman's canoe. The "rake" consisted of a wooden shaft with a row of sharp bone spikes near one end. The small fish (about one foot long) were impaled on the spikes as the fisherman swept the rake through the throng of herring and, in one motion, lifted the fish over the canoe and shook them off into it. The canoe would quickly fill to capacity.

On the Northwest Coast, collecting herring eggs deposited on seaweed had been an age-old practice. However, by the time the Europeans arrived, the Native Americans had developed a much more efficient technique to obtain the delicacy. Just before the great hoard of herring arrived at the spawning areas, the Native Americans covered the bottoms of these bays, inlets, and estuaries with small hemlock branches weighted down with stones. Evidently, the herring found the hemlock boughs a very satisfactory place to deposit their eggs: in a good year the boughs would be covered with a gelatinous mass up to three-quarters of an inch thick. Since the heavy spawning lasted only four or five days, timing was essential. If the boughs were put down too early, the eggs of early arrivals (while too few to warrant collecting) would have begun to develop and their presence among the eggs deposited later would spoil the final product. Once the roe-laden boughs had been collected and brought ashore, the spawn could be eaten right away or preserved by drying. While specific procedures varied, a widely used method was to hang the roe-laden boughs on drying racks. These consisted of a series of horizontal poles one above the other up to fifteen feet high. After drying, the eggs were stripped from the boughs, cleaned, and stored in baskets.

Another fish that spawned in the waters of the Pacific Northwest and played an important role in the lives of Northwest Coast people was the eulachon or candlefish. This small fish spawns in the lower sections of major rivers from the Fraser River north. It is edible but very oily. (So much so that when dried, it will burn if ignited.) Indeed, its oil made the eulachon very important in the Northwest Coast scheme of things. This oil was the favored condiment in most of the region, serving much the same function as butter or ketchup today. Although dip-nets and herring rakes can be and have been used to scoop up the eulachon as they swarm upstream, quite some time before the arrival of the Europeans, a much more efficient device, the bag net, had come into use. According to the traditional accounts recounted above, the idea came from a Haisla girl, and the Tsimshian, whose homeland is the northern coastal area of (mainland) British Columbia, are said to have learned of it from the Haisla. The Tlingit of the Alaska Panhandle, in turn, learned of it from the Tsimshian.

Extracting the oil from eulachon was a major operation. While the details of the process varied from one locality to another, its basic steps were as follows. First, canoe loads of the fish were dumped into a pit and allowed to "ripen," to decompose sufficiently for the oil to be released from the disintegrating tissues. This could take from ten days to three weeks. Then water was heated to boiling in bentwood boxes (see below) by dropping hot rocks into them and the rotten fish were dumped in. As stones cooled, they were removed for reheating and replaced with hot ones. After several hours of simmering, the mixture was left to sit. In time the oil, being less dense, collected at the top and, when it cooled, assumed the consistency of grease and was easily skimmed off. Sometimes those engaged in extracting the eulachon oil settled one of their wooden dugout canoes snugly into the sand, secured it with stakes and cords, filled it with water, and proceeded as they did with the wooden boxes.

While some of the eulachon oil was produced for their own use by those catching and processing the fish, much of it was intended for trade. Although the desire for the product extended throughout the region, ownership of the right to fish where the eulachon made their spawning runs was very restricted. Those people whose territory included stretches of rivers favored by these fish not only were assured of meeting their own needs but also of being enriched through trade. In some cases, they did not catch or process the eulachon themselves but traded the right to do so for goods offered by those who wanted the oil. The latter would then camp by the river, catch the fish, render the oil, and depart with it. Most notable among those favorably situated were the Tsimshian whose territory encompassed the Nass and Skeena Rivers. Eulachon oil from the Nass was recognized, even by those from other oil-producing areas, as the best in the entire region. So prized was the oil that people from Oregon to Alaska converged on the Nass each spring to trade for it.

Among those whose home territory boasted no rivers in which significant numbers of eulachon spawned and who came to trade were the Nuu-chah-nulth of the west coast of Vancouver Island and the Haida of Haida Gwaii. The Haida excelled at wood working, and from the very tall, straight-grained red cedars that grew on Haida Gwaii, they fashioned huge dugout canoes up to fifty feet long and eight feet abeam, which they brought to trade. These imposing craft were very much in demand. The Haida also offered elaborately carved wooden chests, chewing tobacco, and dried halibut. Even the Tlingit, who had some eulachon oil sources of their own, came to trade, offering a variety of goods including shield-like objects of native copper. Although relatively rare, copper does occur naturally in metallic form and was highly esteemed by the Native people of the Pacific Northwest. Rather than for any practical purpose, copper was sought as a symbol of wealth and power.

The Tsimshian also traded eulachon oil for buckskins from people living in the interior of the continent. The path over the mountains by means of which this exchange took place was known as the "grease trail." The hides and pelts of mammals were especially prized by the people of the Northwest Coast because, unlike the prey of big-game hunters (deer, bison, moose, etc.), their primary source of food, fish, provided nothing from which clothing could be made. Indeed, few if any of the highly developed cultures of the Pacific Northwest could have flourished where they did had a solution to this problem not been found.

As noted at the beginning of the chapter, the second major element in the Northwest Coast economy was the hunting of marine mammals. While much less important as a food source than fish, the marine mammals yielded oil, furs, hides, and raw materials for a variety of useful devices. Furthermore, sea-hunting, as well as the hunting of large land mammals, had a central role in the cultural and ceremonial life of the people. Success in hunting was accorded high status and was much celebrated in song, dance, and dramatic performance. In sharp contrast, the "harvesting" of anadromous fish by means of nets and weirs was a low-status activity, even though it provided the bulk of the Natives' diet. Gathering vegetable foods, primarily berries and roots, was also a low-status task, performed by women; such foods made up only a fraction of the Northwest Native Americans' fare. A favorite prey of the Northwest Coast hunters was the sea otter. Its flesh was considered a delicacy, and its luxurious pelt was highly prized for making clothing. Seals were also important, both for their meat and for their hides. After cutting off their heads and feet, the Native Americans peeled the skins of the

seals' back over their bodies without making an incision in either their bellies or their backs. Leaving the hair side in, they secured the orifices of the sealskins and filled them with air before hanging them up to dry. Attached by a line to harpoon heads, these inflated sealskins played a vital role in hunting marine mammals by helping to tire out the harpooned animal by making it more difficult for it to dive and stay submerged, and by revealing its location to its pursuers.

These sealskin floats are but one of several sophisticated products of human inventiveness and skill that were crucial to Northwest Coast whale hunting. Others include the sleek dugout canoe, the toggling harpoon head, the tough but flexible line or rope attached to it, the skill of the rowers, etc. All in all, there could hardly be a more impressive display of what can be achieved when people are freed from having to set out periodically in search of sustenance while also carrying all of their possessions. These accomplishments may seem unremarkable, but keep in mind that they were achieved without the benefit of metal tools or agriculturally produced food or fiber. Since flaked stone tools (described in Chapter III) are not very effective for cutting and shaping wood, the Native Americans could not have fashioned their sturdy dwellings, magnificent dugout canoes, elaborately carved totem poles, exquisite carved wooden chests, etc. had they not devised more efficacious methods of woodworking. (Some of the techniques they developed to meet this and other challenges will be discussed following the account of whale hunting below.)

Only the Nuu-chah-nulth of the western coast of Vancouver Island, their cousins of the adjoining tip of the Olympic peninsula (the Makah), and small neighboring groups engaged in whale hunting. Success in this most spectacular of hunts was the ultimate demonstration of strength, courage, and skill. The whale hunt also exemplifies the hierarchical, kinship-linked socio-economic structure of Northwest Coast peoples. Despite some variations from one group to another, the basic pattern was as follows. The chief was not only the leader of the community but also the wealthiest man in it. He was the only one with the resources to outfit a whale hunt and the inherited right to organize and conduct it. It was also the chief's right and responsibility to assume the harpooner's position in the lead canoe and to initiate the attack on the animal by plunging his harpoon into it. The **harpoon** itself was a massive yew shaft up to 15 feet long, tipped by a sharp-edged, detachable mussel-shell head with two toggled bone barbs, i.e., barbs so mounted that reverse pressure on the harpoon head would cause them to swing out and dig in more firmly. (The excavation at Ozette turned up a number of these harpoon heads, carefully packed in cedar bark pouches to prevent chipping and dulling.) Too heavy to be thrown, this enormous hunting weapon had to be thrust directly into the whale, preferably just behind the left flipper.

Good timing and good fortune were essential if the whalers were to get within range of their prey and yet avoid its powerful tail, which could reduce their fragile craft to splinters in the blink of an eye. Also of vital concern was the direction taken by the whale after it was harpooned. If it headed toward the open ocean rather than toward shore, grave hardship—perhaps disaster—faced the whalers. Since they believed that these immense creatures were the equivalent of canoes for certain spirits, the Indigenous people of North America were anxious to gain the cooperation of these spirits and observed elaborate protocols both in preparing for the hunt and in bringing it to completion. If the first harpoon strike was successful, the hunters waited (at a reasonable distance) for the whale to tire itself out attempting to free itself from the embedded harpoon head and attached floats (see Figure 4-1 below). Since whaling canoes rarely if ever traveled alone, the likely next step was for the harpooner of the second canoe to plunge his shaft into the animal when the opportunity arose, thus attaching another long rope with its sealskin floats. Similar actions continued until the whale was utterly exhausted. The whale hunters then immobilized its flippers and delivered a fatal thrust to its heart with a special lance. They then buckled down to the Herculean task of towing the huge carcass back to their village.

B. Wizardry in Woodcraft

By far the most distinctive and most essential feature of the technology of the Pacific Northwest was woodworking. Without the waterborne transport and the shelter provided by woodcraft, nothing resembling traditional life on the Northwest Coast would have been possible. And while the Native Americans employed a variety of ingenious woodworking techniques, the foundation of Northwest Coast woodworking was a radically new method of making cutting instruments: grinding and polishing rather than flaking and chipping, which had launched human ancestors on the trajectory that has carried their descendants not just beyond the confines of Africa but also beyond the confines of the Earth. Not only are ground stone tools superior to flaked ones for cutting wood, but they are capable of being re-sharpened if dulled or blunted. Of course, this new way of making stone tools was developed at other locations as well as at earlier times, most notably in Southwest Asia some 11,000 years ago. In that case, the transition to polished stone tools was accompanied by (1) a shift from hunting and gathering to farming and herding and (2) the adoption of a sedentary lifestyle. Taken together, these fundamental changes are usually referred to as the "Neolithic Revolution" (see Chapter V). Because of a tendency to view the

Figure 4-1. Makah Whaler. In this 1915 photograph, a Makah whaler holds a harpoon with attached floats.

advent of agriculture as fundamental (and sedentism and other cultural developments, including technological ones, as deriving from it), the term is often used to refer simply to the change from foraging for food to producing it. This usage is unfortunate in that, (1) as seen above, a sedentary lifestyle is possible without agriculture, (2) sedentism fosters technological innovation, one example of which very well may be agriculture itself (see Chapter V), and (3) ground and polished stone tools appear to be associated with sedentism even in the absence of agriculture (see just below).

The archaeological evidence indicates that the first human inhabitants of the Pacific Northwest used flaked stone tools of the type described in Chapter III. But in the latter half of the fifth millennium BCE, ground slate lance and spear points begin to show up in the archaeological record along with a marked increase in bone and antler tools—not just in numbers but in distinct types of tools. **Ground stone celts**, small ones at first, also begin to show up during this same period along with other evidence of woodworking, including bone and antler wedges and ground stone mauls. Shaped like a short chisel or ax-head, celts were ground to a fine edge with sandstone and skillfully mounted on wooden or bone hafts or handles to make adzes and other tools. The blades of adzes are mounted at right angles to their handles, like those of hoes or mattocks; but the handles need not be long. Indeed, an adze with a D-shaped handle, rather like that of a modern hand saw, was used by the Nuu-chah-nulth, the Makah, and other Native peoples living on Vancouver Island and the adjoining coast of British Columbia and the state of Washington. Because **adzes** equipped with these ground stone blades made woodcutting and shaping so much more efficient, large-scale woodworking became possible: trees were cut down, planks were pried from their trunks, and dugout canoes emerged from hallowed out logs. There is archaeological evidence of plank houses and, indeed, villages of them by 1800 BCE, indicating a sedentary or semi-sedentary lifestyle. Thus, Northwest Coast Culture exemplifies a general pattern of ground stone tools being associated with sedentism or semi-sedentism and with fishing and hunting in a maritime context.

In addition to stone, the Native Americans of the Pacific Northwest employed quite a variety of materials—bone, antler, shell, and even teeth—to make blades and points. As noted above, the points or blades of the harpoons used in whale hunts were fashioned from mussel shells. Like slate, which is also hard and strong but fractures into thin plates, mussel shells take less time to work into a blade than would be required if starting with a thick block of material. Indeed, making a point or blade from slate or shell by grinding and polishing is more time-consuming than fashioning a similar item by flaking, but the advantages gained evidently justified the extra effort. The Native workers also employed fire in felling trees and in hollowing out the interior of canoes and large trough-like containers. Using water, they were quite expert at **controlled burning,** which gave them an enormous advantage when it came to cutting down large trees. If undertaken with an adze alone, the process parallels the gnawing of beavers and is laborious in the extreme.

The hafts or handles of tools were often elaborately carved. These images served not only as vehicles of aesthetic expression but also (following the pattern discussed above) as a means of communicating with and enlisting the forces of the spirit world. The extent to which technology and matters of the spirit were intertwined in Pacific Northwest culture is illustrated by the way the Natives went about hunting the beaver and making use of it. From a purely technological point of view, the most impressive thing is the ingenious way the Native Americans mounted beaver teeth in wooden hafts to make chisels. But it would be a mistake to assume that they concerned themselves solely with such factors as the hardness and strength of materials and geometrical configurations. On the contrary, they would have entertained no hopes of thus availing themselves of the wood-cutting powers of the beaver without the assent of its spirit. Not only did the Native American craftspeople recognize they needed the cooperation of the beaver spirit, they tended to view hard, strong chisel blades as manifestations of the power of that spirit.

The cornerstone of Northwest Coast woodcraft was the making of **boards.** Although they had no saws, the Native Americans nevertheless managed to produce planks that were relatively thin and flat by taking advantage of the straight grain of the western red cedar. Starting at the end of a log or at one of two notches cut in a standing tree, they used stone mauls, hand-held or hafted on wooden handles, to hammer chisels and wedges into the log or tree trunk in order to split off long slabs of wood. They then used adzes, chisels, and knives to give each timber its final form. Boards thus produced were up to a meter wide and as much as seven meters long, but only around two inches thick. Finishing touches included grooves in roof planks to guide the runoff from rains and elaborately carved images on planks to be used in chiefs' houses. The frames of the Natives' large, communal dwellings featured heavy wooden posts, set in the ground, supporting massive horizontal beams overhead. Without their covering of roof planks and siding boards, these heavy timbered structures are surprisingly reminiscent of classical Greek and Egyptian temples.

In addition to houses, boards were also used to make a variety of boxlike containers that played indispensable roles in the Northwest Coast way of life. The Native craftspeople fashioned all four sides of each box from a single board by cutting three grooves across it, steaming the board at these places to make it pliable, and bending the board by ninety degrees at each groove so that its two ends met. To make the corner joint thus formed strong and tight, a groove or "rabbet" was cut across one of the two ends of the board and fitted the other into it. The two ends of the board were worked until they fit together so nearly perfectly that, when secured with pegs or withes of spruce root, they formed a watertight joint. To achieve such precise fits, the Native craftspeople employed an ingenious procedure not unlike that followed by modern-day dentists to locate projecting portions of the chewing surfaces of teeth. After fitting together the two surfaces to be joined as exactly as possible guided by the unaided eye, the craftspeople rubbed a mixture of charcoal and grease on one of the surfaces, then brought the second into place against it. The result was to clearly mark the projecting portions of the second surface, which were then chipped away with a chisel. The same techniques were used to fit and secure the bottoms of the boxes. To make a rabbet joint between the bottom and the sides, a notch or "rabbet" was cut at one edge and along the entire length of the board being formed into the sides of a box. A similar notch was cut along all four sides of the board being made into a bottom for the box. When finished, it fit into place as neatly and securely as a champagne cork. There it was made fast with pegs or withes.

Boxes thus made are absolutely watertight, which was perhaps their most important feature. Since the Northwest Coast Natives had neither pottery nor metal pots or containers, they utilized wooden boxes both for cooking and for storing liquids. Food was cooked by dropping it into boxes of boiling (or nearly boiling) water, which had been heated by hot rocks taken from the fire and placed in it. Small square boxes served as water buckets, large squat ones as tubs for storing oil from candlefish and whale blubber. Furnished with tight-fitting lids (rabbeted or shoe-box style), boxes were used to store nearly everything: clothing, furs, blankets, masks, dance whistles, and other ceremonial paraphernalia, as well as tools, equipment, and materials. A box recovered from the old house at Ozette, Washington, contained fishhooks and materials for making them, bits of bone, a whetstone, and fine cord wrapped around some sticks.

The crowning glory of Northwest Coast woodcraft was the **great seagoing canoe**. Except for up-sweeping bow and stern pieces, which were attached in the water-tight fashion described earlier, each of these magnificent vessels was hewn from a single log. Of course, the Native craftspeople produced a variety of watercraft in this manner, the most humble being a small, one-person "shovel-nose" rivercraft. But it was in fashioning large seagoing canoes, some exceeding fifty feet in length and seven feet in the beam, that they demonstrated their ingenuity and consummate skill. Starting with a carefully selected red-cedar log (or section thereof, split from the trunk of a standing tree in the way described earlier), the canoe maker set about shaping the exterior of the hull with adze and chisel.

If a craft is to lie evenly in the water and steer properly (not list or pull to one side), it must be (laterally) symmetric in shape and have sides of the same thickness. Guided only by the unaided eye, the Native craftspeople were able to produce remarkably symmetric hulls. When they had finished carving the exterior surface, they used it as a guide for hollowing out the interior. As noted earlier, they often used fire in the process. To gauge the thickness of the sides, the canoe maker placed one hand on the exterior surface of the hull and the other hand on the interior surface, opposite it, and slid them along the sides of the craft, feeling for bulges. Another method for producing uniformity in the thickness of a hull was to drill holes of a specific depth in the exterior surface, after it had been shaped, and then to hollow out the inside until the holes were reached (these holes were subsequently plugged). Typically, the bottoms of these canoes were about two fingers' width in thickness. The sides thinned progressively to about half this thickness at the gunwales. This configuration made the craft more stable and less vulnerable to puncture.

When the canoe maker had finished his work with the hand-adze, the hull was covered with a regular pattern of cut marks: the smaller and more indistinct, the more highly regarded the quality. Like the powers of the whale hunter, the special skills of the canoe maker were believed to have been bestowed upon him by spirits and to involve, in their exercise, the operation of spirit power. Just as the cutting effectiveness of a beaver-tooth chisel was believed to be spiritual in nature, so were the special abilities of the canoe maker.

To increase the beam of the canoe, and thus both its capacity and stability, the Native craftspeople next used heat and moisture to soften the wooden hull, stretched its sides apart at the top, and secured them in that position with small boards, or thwarts, which also served as seats for the rowers. "**Steaming**" an object of this size is no small task. The canoe makers filled the wooden shells with water, built fires along their sides to warm them, and dropped red-hot rocks into the water. Thus treated, the wooden hulls quickly softened and could be reshaped relatively easily.

In order for the canoes to glide silently through the water and not alarm the animals being hunted, it was essential that their external surfaces be extremely smooth. To accomplish this, the Native workers used sandstone and sharkskin to "sandpaper" the hulls. They also scorched their external surfaces by rubbing them with the burning ends of cedar sheaves. This process serves to burn off splinters and to prevent sun-rents and insect damage. Sometimes the Natives rubbed oil into a canoe's outside surfaces.

The largest canoes were made by the Haida who lived on Haida Gwaii and Prince of Wales Island. These round-bottomed craft had great, soaring prows and high sterns that helped prevent them from being swamped by the swells encountered in the vast open expanses of water surrounding the islands. The large Haida canoes were used in trade, in ceremonial activities, and in warfare. One of them could carry as much as two tons of cargo or forty warriors. Almost as large and of superior nautical design were the canoes produced by the Nuu-chah-nulth (see Figure 4-2 below) of the west coast of Vancouver Island and their cousins, the Makah, of the northwest corner of the Olympic Peninsula. The bottoms of these vessels were flat, and so were the sides, which angled up and out. Their gunwales flared outward to ward off the seas. These canoes also differed from those made by the Haida in having a lower, vertical sternpost and a less massive prow-piece. The prow-piece formed part of a gracefully arching bowline that swept upward from a sharp-edged forefoot. So good are the lines of these canoes that some have seen in them the inspiration for the American clipper ship.

C. Woven Wonders

An absolutely vital element of the Northwest Coast way of life was the production of **fabric**. While a central cultural role for textiles is to be expected among herders and farmers, it is highly unusual among foragers. Two key factors (in addition to climate) account, in large measure, for this exceptional state of affairs: (1) the primary food source of these Native Americans (fish) and (2) their sedentary (or semi-sedentary) way of life. Unlike hunters, whose main source of meat is land animals, the people of the Pacific Northwest had available to them only a relatively small number of pelts and hides. Consequently, the potential for human habitation posed by the abundance of food sources in the Pacific Northwest could be realized only if an alternative to animal skins could be found for making clothing. (While the climate is relatively mild and the Native Americans tolerated cold to a degree that taxes the imagination, clothing and shelter are essential here to the survival of all but the hardiest of individuals.) The solution to this problem, referred to just above, is woven fabric. But where would non-agriculturists obtain the requisite large supply of fiber, and how exactly did the Northwest Coast people (who lived by hunting, fishing, and gathering) take up the art of weaving—or even become aware of that possibility? As it happens, the natural environment did offer abundant sources of fibrous material suitable for making fabric, mats, and cordage. These included the inner bark of red and yellow cedar trees, tough withes from these trees, the slender roots of spruces,

Figure 4-2. Nuu-chah-nulth canoe. This cedar canoe dates to 1930. Image credit: Leoboudv. CC-BY-SA 3.0.

etc. Especially important for making clothing were the strong but satiny fibers of the inner bark of the yellow cedar. However, as the inner bark of cedar trees lies hidden beneath the furrowed, gray, or brownish-gray outer bark, how might the Native Americans have become aware of it?

Actually, given certain of their other activities, they could hardly have failed to do so. First, as seen above, woodworking was a central element of Northwest Coast culture, and red cedar was ideally suited for making boards and canoes, while yellow cedar was the material of choice for wood carving. It is relatively soft yet durable and, unlike red cedar wood, does not split easily. Thus, the Northwest Coast Natives would have been intimately familiar with the characteristics of both red and yellow cedar. Furthermore, they also practiced one of the most universal and ancient of the practical arts: making containers from tree bark. For obvious reasons, leaves and bark were among the first—if not the first—things used to facilitate collecting and transporting fruits, vegetables, nuts, seeds, and the like, very few of which can be utilized in significant quantities without containers of some kind. As they accumulated knowledge and developed skills, people were able to fashion quite a variety of useful items from the bark of trees, not only containers but also such things as canoes and even houses. Although wooden boxes had assumed a central role in the material culture of the Northwest Coast long before the arrival of the Europeans, the Native Americans continued to use sheets of cedar bark in a variety of ways, including the production of containers: from large boxes to pouches. (As noted earlier, the harpoon heads recovered at Ozette had been stored in cedar bark pouches.) Evidently, the Northwest Coast people were quite familiar with the properties of cedar bark through long practice.

They collected cedar bark by pulling strips of it from standing trees, but this was possible only when the sap was running. Furthermore, processing the bark to obtain usable fiber and producing fabric from it were complex, time-consuming operations that simply would not have been possible without a place to store raw materials and projects in various stages of completion as well as a place to work over extended periods of time. The homes of the Native Americans of the Pacific Northwest, like those of most people prior to the Industrial Revolution, served as workshops and warehouses. The old Makah plank house at Ozette was crammed with tools and materials. In addition to boxes and baskets filled with specialized implements and supplies—like the fisherman's kit described above and others (see below)—excavators found rolls of cedar bark, neatly tied bundles of withes and small roots, slabs of whale baleen, lumps of red ocher (for making paint), etc. The hundreds of small cedar splinters littering what was once the floor bear witness to the woodworking of a bygone day. The use of dwellings as workshops and warehouses is, of course, a continuation of that most ancient of human living patterns, the use of base camps for a variety of activities.

The workshop-warehouse dwellings of the Native Americans of the Pacific Northwest represent, from a technological point of view, a major step forward. Not only did they make it possible for the inhabitants to lay up large quantities of useful materials (and to do so at the most advantageous of times), they also allowed the Native Americans to engage in elaborate processing and fabrication procedures that might last for months and would have been impossible for nomads to undertake. More importantly, having a more-or-less permanent place to work and store things enabled the people of the Pacific Northwest to develop a complex, interconnected set of technologies impossible otherwise. Since technological innovation is often a matter of combining pre-existing devices and techniques or deploying them in novel ways, the larger the technological repertoire of a people, the more likely it is to become even larger and more powerful. (Manifestations of this basic principle of technological dynamics occur repeatedly throughout this book.) Consequently, the sedentary or semi-sedentary way of life of the Native Americans of the Pacific Northwest not only allowed them to add to their tool kit whenever the opportunity arose but also greatly enhanced the probability that new technologies would indeed emerge. From a technological point of view, Northwest Coast **weaving** is particularly noteworthy because it illustrates so well both the opportunities afforded by sedentism and the emergence of one craft from another. Let us now sketch the basic elements of Pacific Northwest **fabric production** and its origins in **basket-making.**

Gathering and processing cedar bark was a major part of the Northwest Coast way of life. In spring, when the rising sap made it easy to pull long strips of bark from standing trees, the Native Americans ventured into the forests and picked out relatively young cedars with smooth expanses of bark stretching upwards to a considerable height uninterrupted by limbs or branches. The bark gatherer first made a horizontal cut through the bark near the ground and pried the bark loose from the tree at the cut with a sharp instrument, splitting it upwards at each end of the cut. Continuing to work in this way as high as she or he could reach, the bark gatherer peeled a wide strip of bark from the tree. Then the bark gatherer grasped the end of the strip in both hands and backed away from the tree, tugging and jockeying as the bark came loose, higher and higher, until it tore free. The next step was to separate the inner bark from the outer bark, which was accomplished by prying them apart at one end of the strip and pulling them away from each other. The strips of inner bark were then folded into bolts for transport back to the village.

As a source of fiber for making clothing, cedar bark has advantages as well as disadvantages when compared to both cotton and linen. Picking individual cotton bolls and removing the seeds is very labor-intensive, and this severely limited the usefulness of cotton until mechanical devices, especially the cotton gin, were introduced to handle these tasks. While not as intractable as cotton, obtaining linen fiber from flax poses its own formidable challenges. Both cedar-bark fiber and flax fiber are obtained from the outer layers of the stems of plants (in the case of cedar, from the bark). Obtaining usable fiber from such sources requires, firstly, separating those fibers from the rest of the plant. In the case of cedar-bark fiber, this is accomplished simply by stripping the bark from the tree and pulling the inner and outer bark apart. The process of separating the desirable flax fiber from the rest of the plant is more complicated. It involves, first of all, "retting": soaking the plants in ponds, streams, or tubs or letting them lie in the field subject to rain and dew, which softens the gums that bind the fibers of the stalk together and expedites the decay of the non-fibrous parts. Then, in a series of steps involving bending, beating, and combing, the bonds that hold the components of the stalk together are loosened and the fragments of brittle parts are removed, leaving the durable bast fibers to be spun into thread.

Although the Northwest Coast people were able to separate—quite expeditiously—the inner bark of cedar trees from the rest of these giant plants, they had to devote long hours to the laborious tasks of either converting the cedar bark strips into fluffy strands of fiber or splitting the layers of cedar bark fibers apart (to obtain thinner and more flexible strips of the material) and dividing the thin strips thus obtained into narrower ribbons suitable for making baskets, rope, or string. To accomplish these tasks, the Northwest Coast women had developed ways of loosening the bonds between the layers of cedar bark fibers and between the individual fibers. These techniques were very similar to those employed in working with flax: **soaking, bending, and beating**. The tools used and the specific steps taken varied from one Northwest Coast society to another, and they depended upon whether red cedar bark or yellow cedar bark was being processed and upon the nature of the intended product. In the case of red cedar bark, the initial steps were quite straightforward: after drying the strips of bark, the Native American women refolded them, packed them in boxes or baskets, and stored them in their homes until the winter months, during which most of the work of processing the cedar bark and fashioning useful items from it would be done. Yellow cedar bark is tougher and was preferable for a variety of textile applications, but processing it was more complicated. After a longer initial drying period, yellow cedar bark was soaked in water for several days, then pounded with a whale bone bark beater or a stone, and dried (again) before being stored. To transform the stiff strips of dried inner cedar bark into fluffy skeins of silken fibers, the women of the Northwest Coast subjected them simultaneously to bending and beating. They pounded the strips of cedar bark with a relatively blunt instrument as they were being pulled over the edge of a wooden block or board or across a groove in a flat surface. This process loosened the bonds between the layers of fiber as well as those between the individual fibers; hence, it is called "shredding" and the instruments used to beat the bark are called "cedar bark shredders," even though no cutting (such as that which takes place in modern paper shredders) was involved. While the inner bark of the red cedar was shredded while dry, that of the yellow cedar was repeatedly soaked and then beaten to soften it. If greater fluffiness was desired, the shredded cedar bark was soaked and then gently beaten with a small stone. After it had dried, the women twisted bunches of the shredded cedar bark back-and-forth with their hands.

The separation of layers and strands of inner cedar bark brought about by shredding is random and uneven, which makes shredded cedar bark unsuitable for actual weaving. Long, even strands of fiber are required for that, as well as for making baskets, mats, hats, fishing lines, etc. However, obtaining even strands of fiber from cedar bark, making thread with them, and weaving cloth with that thread is far too time-consuming to serve any practical purpose. The solution to this dilemma worked out by Northwest Coast women was to develop a way of binding strands of shredded cedar bark into warm, water-repellant fabric that could be made in much less time than woven fabric. Since this process also requires thread or twine (though much less of it), the method used by the women to make these is outlined immediately below, followed by a delineation of the special features of their alternative way of making fabric. A summary of the basic characteristics of weaving (in the usual sense) is included for comparison.

To obtain the needed long, even strands of cedar bark fiber, the Northwest Coast women followed a procedure quite different from shredding: they repeatedly split the strips of cedar bark by bending them and working them with their fingers until suitable strands of fiber were wrested from the inner bark. Success at this endeavor required both manual skill and discriminating judgment in selecting cedar trees from which to pull bark. From the long, even strands of fiber thus produced, they **made string or twine by twisting two or three strands together.** As have so many around the world since early times, the Northwest Coast women twisted the strands of fiber together either (i) by holding one end of them in one hand while rolling them on one thigh with the flat of the other hand or (ii) by twirling them on a hand spindle.

Once thread or string is available, **weaving** becomes possible. Weaving cloth thus constitutes the realization of a potential inherent in the art or practice of making string or thread. While there are many specific ways of doing it, they all share certain basic characteristics. One is the "warp," a series of (more or less) parallel threads, suspended either vertically or horizontally by means of a frame called a "**loom.**" In the simplest case, weaving consists of passing a thread (referred to as "weft" or "woof" thread) alternately in front of or behind (over or under) each of the warp threads as the weaver works her way from one edge of the warp to the other. Each time this process is completed, the weaver simply repeats it just below (adjacent to) the preceding weft thread—but this time passing the weft thread in front of (over) those wrap threads that the preceding weft thread had passed behind (under) and vice versa. A special feature of fabric production on the Northwest Coast was weaving with two weft threads. In this case, as the weaver works her way across the warp, she passes one weft thread in front of each warp thread while at the same time passing the other weft thread behind it. Proceeding on, she crosses the two weft threads before catching the next warp thread between them so that the one that was in front of the warp is now behind it and vice versa. (The process thus resembles sewing with lock stitches.) Furthermore, as the weaver proceeds across the warp, she crosses the two weft threads in such a way that first one and then the other is on top. As a result, the two weft threads are wound or twined around each other. The result is a very tight and sturdy weave. Fabric so woven is said to be "twined." **Twining** is one of the most basic, most widely used, and oldest methods of weaving. It was used by the Tlingit women of the Alaska panhandle to make the superb spruce-root baskets for which they are so well known. Moreover, it is from this basket-making tradition that the weaving of Raven's Tail and Chilkat robes seems to have emerged, as will be discussed below.

Since each weft thread (or pair of weft threads, in the case of twining) is placed in close contact with the previous one, the length of a piece of cloth being woven increases by only one diameter of the thread being used each time the weaver works her way across the warp. Evidently, making fabric in this way requires an enormous amount of time. If, in addition, one considers the time it would take to make the warp threads as well as the weft threads from yellow cedar inner bark, it becomes evident that it would have been impossible for the people of the Northwest Coast to meet their everyday needs for clothing, blankets, etc. using fabric woven from cedar bark thread alone. Some (actual) weaving was done using thread made of a combination of cedar bark and other kinds of fiber, but only the very wealthy (primarily chiefs) could afford items thus made and wore them only on special occasions. On the other hand, the Coast Salish and other Native Americans of western Washington and the southern coastal areas of British Columbia did engage extensively in actual weaving, although they used quite different materials and techniques, as discussed below.

Aside from these groups, the people of the Northwest Coast took advantage of the long fibers of inner cedar bark which enabled them to produce very serviceable fabric in much less time than actual weaving would have required. They began the process by attaching small bundles of shredded cedar bark evenly along a horizontal bar supported at each end by a vertical post. Such devices are called "**half-looms**" or "**single-bar looms.**" The woman making the garment, blanket, or rug then proceeded to bind this expanse of cedar-bark fiber (hanging from the horizontal bar and serving as a warp) into a solid fabric by means of "twined" pairs of weft threads woven horizontally across it at one-half-inch intervals. Just as in twined weaving, she worked with one weft thread on each side (front and back) of the warp; but in this case, the pairs of weft threads do not catch individual warp threads (as in actual weaving, described above) but rather small bundles of shredded cedar bark roughly the thickness of a pencil. This alternative method of making cedar bark fabric also differs from actual weaving in that it leaves space (about one half inch) between the pairs of twined weft threads running horizontally across the warp. While the fabric thus produced is not as tight or firm as woven cloth, it is nonetheless quite warm, due to the fluffiness of the cedar-bark fiber, and is softer and more flexible than woven fabric. And, perhaps most importantly, it does not take nearly as long to make. Not only is the labor of preparing a warp comprising individual threads avoided, but because space is left between the rows of twined weft threads, much less (weft) thread is needed and much less labor is required to bind the shredded cedar bark into fabric.

A cedar-bark robe or blanket could be made more elegant and more desirable by fastening a small strip of fur around the edges or by inserting an occasional weft thread of mountain-goat wool. The more numerous and the more closely spaced the wool weft strands, the more highly prized—and the more expensive—was the blanket or robe. Only a very wealthy person, such as a chief, could afford one of these items with most (or all) of its weft threads made from goat wool or from some other scarce and expensive material. Another method used to enhance the quality of cedar-bark textiles was to work duck down into the bundles of warp fiber before the weaving was done. Because of the fluffiness of the down, the widely spaced cedar-bark weft threads are not visible in articles thus made. As a result, such items are warm but remarkably light and have the overall look of mallard down, a soft rich brown.

Utilizing a warp composed of individual threads opens up a whole new range of possibilities as regards both materials and methods of fabrication and thus of the appearance of the woven product. A striking example is afforded by the practice of inserting at regular intervals warp threads of a different material from the rest. This process is analogous to using different kinds of weft threads, as just described. In a particularly ingenious variation of this technique, narrow strips of sea-otter pelt were introduced between each pair of twisted cedar-bark warp threads, with the fur sides all facing in the same direction. On that side of the tightly woven finished product, very little can be seen other than the deep, sea-otter fur. The reverse side is dominated by the twined, mountain-goat-wool weft which, in a surviving museum specimen, catches the cedar-bark warp threads two at a time (see discussion of twill, below). The net result is a blanket that is fur on one side and (woven) wool on the other!

The combination of different kinds and colors of warp as well as weft threads and different weaving techniques (different "weaves") makes possible a great variety of textures, colors, and patterns of woven fabric. The way a fabric is woven determines which thread—and hence which color—is visible at a given spot. The Tlingit weavers of the Chilkat region exploited this fact expertly in making the celebrated robes or blankets named for the area. To produce their elaborate designs, they used threads of four different colors: white (the natural color of carefully washed wool) and black, blue, and yellow (colors achieved by dying). The use of different colors of thread in both warp and weft produces such widely produced patterns as plaid. This possibility was also exploited by the weavers of the Pacific Northwest: the remains of a small plaid blanket were found during the Ozette excavations.

Certain basic weaving patterns seem to have found their way into most textile-making traditions. One of these is **twill**, in which each weft thread (repeatedly) passes over two warp threads at a time and each successive weft thread splits the pairs of warp threads taken together by the previous one. When a weft thread crosses over two warp threads, a tiny raised place on the surface of the fabric is produced. Since each of these raised places is displaced by half its length relative to those produced by the previous weft thread, fabric thus woven has a handsome, ribbed surface with its parallel ridges running diagonally. Denim for making jeans and khaki for military uniforms have a twilled weave—and so do the Chilkat robes! Specifically, the Chilkat weaver used a "twilled twining" technique: working with two weft threads simultaneously, one in each side of the warp, in the manner described earlier, she caught two warp threads in each of the "lock stitches" she made as she worked her way across the warp. In the next trip across, each of the pairs of warp threads thus bound together was split, with each member of the pair now bound to the warp thread on its opposite side. The makers of the Raven's Tail robes also employed the twilled twining technique in addition to a number of other weaving patterns.

If weft threads of two different colors are used when twining, the possibilities for creating geometrical patterns in fabric quickly become evident. In each trip across the warp, first one color and then the other will show itself on a given side of the fabric. If, in succeeding rows, the color of the first "lock stitch" alternates, the fabric or basket will have a fine checkered appearance. The Tlingit designation for this pattern, which they used in basketry, is "Shuck kukh" ("strawberry-basket"). On the other hand, if each weft row begins with the same color, the fabric will be covered with narrow vertical lines of alternating colors.

Horizontally adjacent "lock stitches" of the same color can be produced by utilizing a "double twist"; instead of crossing the weft threads between a given pair of warp threads, the weaver twists them around each other so that each weft thread remains on the same side of the warp in successive "lock stitches" or weaving segments. An uninterrupted horizontal line of a single color can be made in the fabric by repeating this process again and again. And by employing this technique in successive weft rows, the weaver can create an expanse of uniform color. Hence one can easily produce vertical or horizontal lines of one color against a background of the other color. Diagonal lines (or changes from one color to the other along a diagonal) can be created by arranging for changes in color to take place at successively earlier (or later) points in each weft row. In this context, it is no surprise to find Northwest Coast baskets and blankets decorated with geometrical patterns. Figure 4-3 below shows a pair of Tlingit baskets.

Figure 4-3. Tlingit *Kadádzaa yéit* (berrying baskets), c. 1900. Image credit: Joe Mabel. CC-BY-SA 3.0

The finest example of geometrically patterned, twined textiles in the Pacific Northwest is the Raven's Tail dancing robe. Typically, these rectangular cloaks are covered with a grid of geometrical designs. Unfortunately, very few of them have been preserved and very little is known about who made them. Indeed, if it were not for a very small number of museum specimens (some of them mere fragments) and the drawings and paintings of a few (mostly European) sailors and adventurers, this art form would have slipped into oblivion. These exquisite garments were produced in the northern part of the Northwest Coast region until roughly the beginning of the nineteenth century. At that time, the Native American weavers apparently shifted their textile production to Chilkat Dancing Blankets (so designated because production of these cloaks was—and remains—centered in the Chilkat region of the Tlingit). Until very recently it was not known that the weaving of these celebrated robes was itself the outgrowth of an older but equally distinguished weaving tradition about which very little is known. However, because of (1) the very ancient and near-universal character of basket-making, (2) the utilization of the same weaving techniques (as noted earlier) and some of the same materials (cedar bark) in making both baskets and robes, and (3) the presence in both types of articles of a number of the same geometrical patterns, produced in the same way, there can be little doubt that weaving in the Pacific Northwest emerged as an offshoot of **basket-making.** Because they were highly skilled in a kind of basket-making that lends itself to a certain type of fabric production, the women of the Pacific Northwest were able to take advantage of the opportunity—and the need—to engage in such an enterprise when the time was right. Their impressive achievements in the textile field provide a powerful illustration of the principle that old technologies are the fertile soil from which new ones spring.

A key step in the emergence of Western science and philosophy was the development in Ancient Greece of deductive proofs of various geometrical relationships, especially among the properties of geometrical figures such as triangles, squares, rectangles, etc. The Babylonians and Ancient Egyptians (and probably others) knew of some of these relationships and made use of them in their large-scale construction projects and in surveying land. Indeed, the term "geometry" derives from the Greek *geometrein*, to measure land (earth). Accordingly, the tendency has been to assume that it was in the context of building and surveying that people not only learned of these relationships but also first became acquainted with such geometrical figures as rectangles, squares, right triangles, isosceles triangles, etc., which are rarely (if ever) observed in nature (prior to the invention of the magnifying glass and other instruments).

However, recent discoveries pushing the date of the earliest known example of weaving back to about 26,000 years ago raise the possibility—indeed, the probability—that human beings first became aware of these geometrical forms in the context of weaving or something very like it. Evidence that people were already accomplished weavers by this early date is provided by impressions on fragments of fired clay recently recovered at an archaeological site in Moravia (a region of the Czech Republic). Not only do the impressions on these clay fragments exhibit the unmistakable pattern of woven fibers, but a closer examination reveals that the weaving technique involved was twining: in some cases simple (or plain) twining, in others twilled twining. These are, of course, precisely the two types of twining that loomed so large in Northwest Coast weaving. Given the way twining lends itself to weaving geometrical designs (as explained above), it is highly likely that woven geometrical designs existed at this very early date. In contrast, early dwellings and other structures were circular or roughly elliptical. By 7000 BCE, evidence of rectangular buildings begins to show up in the archaeological record. Large-scale buildings and engineering projects, of course, came even later.

Although West Coast weaving had its roots in the twining techniques of basket-making and made extensive use of materials (notably cedar bark) that were used in basketry, it was by no means confined to such material resources or methods of fabric production. Indeed, the Coast Salish, who were the premier textile workers of the entire region, made little use of cedar bark fiber and only limited use of twining. They availed themselves of one of the oldest technologies in the human repertoire: spinning relatively short, weak fibers into long, strong threads (discussed in Chapter III). Moreover, they used quite a variety of types of fiber: fireweed and cattail cotton, the down of geese and ducks, and the wool not only of mountain goats but of a breed of small dogs, which seem to have been kept for just this purpose. The Salish artisans beat these ingredients together into a fluffy mass and then spun them into a thick yarn on spindles with large, decorated whorls. Since most of these materials were available only at certain times of the year (and some only very briefly), a protected storage place and workspace was absolutely essential for enterprises of this kind.

Another special feature of Salish fabric production was the use of "**full**" or "**double-bar**" **looms**, which have a second horizontal bar below the first. These looms greatly facilitate weaving with a single weft thread. The warp of the fabric being woven is wrapped around the two horizontal bars, forming a broad, endless belt. The two ends of each of the warp threads are attached to each other by means of a single horizontal string running through loops at the ends of each of those threads. When the weaving is finished, this string is removed and the ends of the blanket or rug separate, with the loops left to form a fringe along each end. Three complete double-bar looms, and fragments of others, were found at

Ozette in the ruins of Makah plank houses. The presence of so many looms in one house, even though it sheltered several families, testifies to the central role of woven materials in the lives of the Northwest Coast Natives.

Coast Salish weavers, working with warp stretched on their two-bar looms, were able to weave with a single weft thread. They employed a twilled checker technique, taking the weft thread over two warp threads and then under the next one. An important feature of twill is that little other than the weft threads are visible. (Strictly speaking, this is true for checkered twill only if it is viewed on the "right" side.) Consequently, the weaver can control the color of the fabric by selecting the color of the weft thread. While it is believed that Salish weavers produced elaborate, colorful designs in their blankets, none antedating the arrival of the Europeans has survived; hence one can only speculate about the specifics of these designs. From the accounts and sketches of adventurers and traders, as well as from surviving samples, we do know that the Salish made **plain-weave** textiles for everyday use. Among the Coast Salish, fabric production reached such a high level of proficiency that woven cloth was cheap enough to be put to common use.

At this juncture one might be tempted to think that Northwest Coast basketry is worthy of note only because it gave rise to weaving. After all, basket-making is one of the oldest and widely pursued of the practical arts. There are at least four reasons to resist this temptation. First, Northwest Coast basket-making is itself an excellent example of the advantages of having a more or less permanent residence that also serves as a warehouse and workshop. Second, basketry, mat-making, and wickerwork were an integral part of the complex system of interconnected technologies upon which the Pacific Northwest way of life depended. Third, if weaving (with its geometrical patterns) did derive from basket-making and related crafts, then the presence of these same geometrical designs on the baskets (and related products) surely signals a much earlier date (than even that of the invention of weaving) for human beings' first acquaintance with geometrical figures like rectangles and triangles. And fourth, Northwest Coast basketry and related crafts are an excellent example of technology providing an opportunity for aesthetic expression.

Although basketry, mat-making, and wickerwork were not unique to the Northwest Coast way of life (in the way that their woodworking and fabric production were), the Native Americans of the Pacific Northwest nevertheless managed to set standards of excellence in these areas as well. Here again they were able to bring together just the right combination of suitable raw material and consummate skill. They used quite a variety of materials as well as weaving techniques, but with twining (the technique they came to use in weaving cloth) playing a central role. Taking full advantage of the toughness and pliability of such materials as spruce roots and split cedar bark, the women of the Northwest Coast wove baskets so skillfully and so closely that they were watertight! The Tlingit even used them for cooking—using the dropping-hot-rocks-in-water technique described earlier. The women also made large, conical rain hats in the same fashion and from the same materials.

The Native American artisans used an impressive variety of methods to decorate their baskets and hats: they painted designs on them; they dyed some of the weft elements so that patterns and designs could be woven into the wicker; they worked lengths of tough grasses and sedges, variously colored, into the wicker as it was being woven; and they employed different weaving techniques to produce an assortment of textures and patterns. Of course, utilizing some of these methods of decoration requires that certain materials be collected at a specific time of year. For example, if a tough swamp grass, often used for such purposes, is to be of the right quality, it must be gathered in August. (In July, for example, its blades are too soft and narrow.) Being able to collect materials—even if bulky and heavy—at the optimum time and store them until needed is seen, once again, as a prerequisite for key practices of the Northwest Coast way of life.

The women of the Pacific Northwest made and used quite a variety of baskets—from the very small to the extremely large, from the very plain and simple to the ornate. Indeed, baskets were the practical, all-purpose container of the region's Native Americans. They used baskets to collect and carry food and material, the most fundamental activity of the hunter-gatherer mode of existence. They used withes and boughs to make baskets that were stiff enough to retain their shape when set down or when carried on the back. A tumpline across the forehead, stretching back over the shoulders, supported the large carrying baskets they employed to gather firewood, cedar bark, clams, etc. This arrangement leaves the hands free for other pursuits. One large carrying basket unearthed at Ozette was still full of rolls of (apparently) freshly collected cedar bark. Indeed, the old house was filled with baskets of all kinds and descriptions. Some were of a type used in gathering clams. These had a coarse, open weave that allowed sand and water to filter out. Also found in the old house were large open-weave storage baskets. These were used throughout the region to store such things as dried salmon, which require ventilation. The Native Americans used small, open baskets for berry-picking and for holding trinkets.

One small basket from the excavation at Ozette appears to have been a weaver's kit. Its contents included a hardwood spindle whorl, two combs, several awls, a whetstone, a bundle of bird bones (probably material for making awls), stone blades, a lump of red pigment, and clumps of fiber (probably intended for making thread). The Native artisans also made small, flat pouches that could be carried at the waist. One of these, found at Ozette, contained wooden fishhooks with bone points, bits of bone suitable for making more points, a whetstone for sharpening them, and a hank of thread. A somewhat larger flat pouch found there held a mussel-shell harpoon point carefully wrapped in cedar-bark fiber to protect its sharp edge, as noted above. These flat pouches and many of the utility baskets were woven of cedar bark in a simple checkerboard pattern, with each weft element repeatedly passing over one warp element then under the next (and with each succeeding weft element reversing the sequence, so that it passed under the warps which the preceding weft had passed over, and vice versa).

This same weaving pattern was followed in making mats of cedar bark, reeds, and rushes. Woven mats played a central role in the daily lives of the Natives of the Pacific Northwest. People slept on them, sat on them, and set out their food upon them at mealtime. They also wrapped the bodies of the deceased in cedar bark mats in preparation for burial. Expertise in wickerwork was also applied in the construction of fish traps. To harvest the great quantities of fish upon which their lives depended, the Native Americans of the Pacific Northwest constructed elaborate weirs and traps. They built fences stretching from riverbank to riverbank, piled rocks to divert the flow of streams and the runs of fish, and fashioned wickerwork fish traps up to twenty feet long and five feet in diameter.

* * *

At the core of the human story is a series of technological innovations followed by fundamental changes in the way people have lived. But it has not always been clear just how these innovations and changes in living patterns are related to each other. It was once assumed that **sedentism** was a consequence and/or a concomitant of the advent of agriculture. Recent studies, however, indicate that agriculture developed among people who were already leading a sedentary or semi-sedentary way of life (as is discussed in the next chapter). Thus there arises the need to consider the significance of sedentism (for technology) separately from matters related to agriculture. Because of their isolation from agriculturalists and of the availability of so much information about their sedentary or semi-sedentary way of life before the arrival of the Europeans, the above case study of the Native Americans of the Pacific Northwest is an excellent forum for reflections on this topic.

Sources / Further Reading

Ames, Kenneth M. Ames, and Herbert D.G. Maschner. *People of the Northwest Coast: Their Archaeology and Prehistory*. Thames and Hudson, Ltd., 2000.

Colson, Elizabeth. *The Makah Indians: A study of an Indian Tribe in Modern American Society*. Manchester University Press, 1953.

Cutter, Donald C. *Malaspina & Galiano: Spanish Voyages to the Northwest Coast, 1791 & 1792*. University of Washington Press, 1991.

Dockstader, Frederick J. *Weaving Arts of the North American Indian*, revised ed. HarperCollins, 1993.

Drucker, Philip. *Cultures of the North Pacific Coast*. Chandler Publishing Company, 1965.

——. *Indians of the Northwest Coast*. American Museum of Natural History, 1955.

——. *The Northern and Central Nootkan Tribes*. Smithsonian Institution, Bureau of American Ethnology, Bulletin 144, 1951.

Emmons, George T. *The Basketry of the Tlingit and the Chilkat Blanket*. Sitka, Alaska: Friends of the Sheldon Jackson Museum, 1993. Reprinted from the *Memoirs of the American Museum of Natural History*, vol. III, part II, July 1903, pp. 229–77, and vol. III, part IV, Dec. 1907, pp. 329–400.

Gunther, Erna. *Indian Life on the Northwest Coast of North America: As Seen by the Early Explorers and Fur Traders during the Last Decades of the Eighteenth Century*. University of Chicago Press, 1972.

Hays, Hoffman R. *Children of the Raven: The Seven Indian Nations of the Northwest Coast*. McGraw-Hill Book Company, 1975.

James, George Wharton. *Indian Basketry*. 1909. Dover Publications, Inc., 1972.

Jewitt, John R. *The Adventures and Sufferings of John R. Jewitt, Captive of Maquinna*. Annotated and illustrated by Hilary Stewart. University of Washington Press, 1987.

Kirk, Ruth. *Tradition and Change on the Northwest Coast: The Makah, Nuu-chah-nulth, Southern Kwakiutl, and Nuxalk*. University of Washington Press, 1986.

Krause, Aurel. *The Tlingit Indians: Results of a Trip to the Northwest Coast of American and the Bering Straits*. Translation by Erna Gunther of *Die Tlinkit-Indianer* (Jena, Germany, 1885). University of Washington Press, 1956.

Moziño, José Mariano. *Noticias de Nutka: An Account of Nootka Sound in 1792*. Translated and edited by Irish H. Wilson Engstrand. University of Washington Press, 1991.

Samuel, Cheryl. *The Raven's Tail*. University of British Columbia Press, 1987.

——. *The Chilkat Dancing Blanket*. University of Oklahoma Press, 1989.

Sproat, Gilbert Malcolm. *The Nooka: Scenes and Studies of Savage Life*. 1868. Edited and annotated by Charles Lillard. Victoria: Sono Nis Press, 1987.

Stewart, Hilary. *Cedar: Tree of Life to the Northwest Coast Indians.* University of Washington Press, 1984.

——. *Indian Fishing: Early methods on the Northwest Coast.* University of Washington Press, 1977.

——. *Stone, Bone, Antler, & Shell: Artifacts of the Northwest Coast.* University of Washington Press, 1996.

Swan, James G. *The Indians of Cape Flattery, At the Entrance to the Strait of Fuca, Washington Territory.* 1869. The Shorey Bookstore, 1972.

——. *The Northwest Coast: Or, Three Years' Residence in Washington Territory.* 1857. University of Washington Press, 1972.

Woodcock, George. *Peoples of the Coast: The Indians of the Pacific Northwest.* Hurtig Publishers, 1977.

Chapter V:
Agriculture, Population Growth, and Linguistic Imperialism

THE SHIFT FROM FORAGING for food to growing crops and herding animals marks a fundamental change in the way people live, one of the most significant that has ever taken place. Not only does the practice of agriculture impose exacting conditions upon those who follow it, but it also paves the way for yet more radical changes in the very nature of human existence. Although this transition is often referred to as **"The Agricultural Revolution,"** it did not take place all at once but in fits and starts over many centuries. Furthermore, it was not one event but many, with several distinct beginnings and numerous strands of development. To make matters even more confusing, the dates of even the key developments continue to be debated. Unfortunately, as pointed out in the Introduction, a formidable barrier stands in the way of the modern reader's understanding of how this transition came about: the presumption that the people who first came to rely upon agriculture as their primary source of food did so to secure for themselves the benefits now believed to derive from the practice of agriculture. As obvious as these advantages now seem, people did not and could not have followed such a course of action for reasons of this kind because the earliest farmers could not possibly have anticipated either the benefits that advances in agricultural would bring or the indirect benefits that the agricultural way of life would make possible. The nub of the matter is that primitive agriculture provides neither a more reliable nor a more nutritious food supply than that typically available to members of hunter-gatherer societies. Early farmers had to accept a more restricted diet, the risks of crop failure (due to adverse weather conditions, pests, plant diseases, etc.), and an average work day that was not only twice as long as the typical hunter-gatherer's but consisted of sheer drudgery.

Primitive agriculture is not the highly efficient enterprise now taken for granted in modern industrialized countries, nor is it even in the same class with that which nourished ancient Babylon, Zhou-dynasty China, or the Maya of the Classic Period. The practice of agriculture is an extremely complex activity requiring a suitable environment, extensive knowledge, sophisticated skills (needed not only in the activity itself but also in the production of the tools necessary for it), and a full complement of plants and/or animals that lend themselves to agricultural exploitation. One could hardly expect all the elements of this complex to have been assembled all at once and for the practice of agriculture to have begun following some kind of "Eureka" moment in which the relevant possibilities were recognized. It takes a long time to accumulate the knowledge, develop the skills, and domesticate the plants and animals needed for an agricultural way of life.

Indeed, it appears that up to two thousand years may have elapsed between the time that edible plants first began to be cultivated and the time when farming became people's principle means of obtaining sustenance. Thus, for thousands of years—right up to the present in some places—people have sustained themselves by one *combination* or another of hunting, fishing, gathering, planting, and herding. The reason(s) for the occurrence of each of the many small steps that culminated in one or another fully developed agricultural system would have been highly dependent upon local conditions at the time. Consequently, there is no simple answer to the question of why people shifted from foraging to farming.

Following a review of the basic features of the genesis of agriculture, fleshed out with two case studies, this chapter focuses upon important effects of the adoption of the agricultural way of life: mushrooming farming populations expanding into larger and larger areas, carrying their language and culture with them, and absorbing or displacing the more dispersed inhabitants encountered. Lastly, attention is directed to the complex connections among technology, language, and other elements of culture. Especially significant is the link between agriculture and **sedentism** and the resulting enhanced opportunity for technological innovation.

A. Basic Features of Transitions to Agriculture

Even so, certain features of such transitions are pretty clear. First of all, as noted in the Introduction, those who initiated the practice of planting seeds in order to augment what nature provides were already leading a sedentary of semi-sedentary way of life. Since initial efforts of this kind were unlikely to produce more than a small fraction of people's food supply, and since realizing the benefits of such efforts requires remaining near or returning to the site, those living under conditions allowing them to maintain a more or less permanent place of residence are able not only to take advantage of opportunities to augment nature's bounty but also to store the fruits of their labor until needed. Thus, it is hardly surprising that the first people known to have domesticated plants had long been engaged in harvesting and storing wild grain in substantial structures as part of their sedentary pattern of living. Indeed, given (i) the capacity of human beings to undertake modifications and interventions in the natural order and (ii) their intimate familiarity with the life cycles of desirable plants and the conditions that affect their growth, it was only a matter of time until someone tried planting the seeds of one or another of these plants.

Taking up the practice of planting seeds rather than just leaving things up to nature is but the first step on the road to the agricultural way of life. A second essential element is the **domestication** of plants and animals, a process involving changes in their genetic makeup manifested in significant differences between the domestic and wild varieties. Some of these have resulted from deliberate human actions (e.g., selecting seeds on the basis of desirable traits); others have not. Characteristically, the edible parts of domesticated plants are larger than those of wild ones, a development almost certainly the result of deliberate human choice. For example, the grains of domesticated emmer wheat are larger and plumper than those of the wild ancestral variety. Similarly, even the earliest known species of domesticated corn (maize) had ears close to two inches long with eight rows of kernels, from six to nine to a row. Later varieties have increasingly larger ears with more rows of kernels. In contrast, teosinte, the ancestor of maize, has only small spikes with a single row of kernels. The domesticated descendants of wild emmer wheat and teosinte share another characteristic: they no longer develop in such a way that their seeds are released and scattered when ripe. In the case of the ancestral plants, the stem-like structures by which the seeds are attached to the plant become brittle and shatter when the seeds ripen. But this does not happen in the domesticated varieties. The ripening grains in the heads of wheat remain firmly attached, as do the kernels on the corn cobs. Plants thus unable to disperse their seeds stand very little chance of surviving in the wild.

Their continuing to reappear in succeeding growing seasons is dependent on the intervention of human beings: their harvest of the ripened grain and storage of some of it in anticipation of the next planting season.

Ripened seeds that remain firmly attached are an advantage when grain is being harvested, since a much smaller fraction of it drops free and is lost during the process. But it does not necessarily follow from the presence of firmly attached seeds that deliberate selection is responsible. As long as a fraction of the seeds in a given crop is weakly attached, an even smaller percentage of those that are harvested, stored, and available for planting the next season will be weakly attached (because a higher percentage of them will have been lost). Repetition of this pattern year after year and century after century inevitably results in the elimination of the mechanism by means of which the wild variety propagates itself and in the completion of the conversion of the plant to a domesticated crop. These changes are also an aid to anthropologists and archaeologists seeking to determine whether the inhabitants of a given site engaged in growing crops. Thus, there are clear indications that domesticated varieties of barley, emmer wheat, and einkorn wheat were being cultivated in a narrow strip of land stretching from the Damascus Basin to the lower Jordan Valley some 9,700 years ago.

The domestication of animals poses especially formidable challenges, requiring, first of all, that human beings gain control, to one degree or another, of the location and movements of the animals. This may involve confining them to a pen or other enclosure, herding them, or guiding their movements in a certain direction. It follows that the people exercising this control must assume responsibility for the animals' food supply. And, if a domesticated population is to be established, they must make sure that provision is made for the animals' reproduction. Animal species vary widely with regard to their suitability for domestication. Among the most unsuitable are those, like antelopes and gazelles, which are adapted to escape swift predators. They are skittish and very difficult to confine, in no small part because they tend to panic when threatened with confinement. Furthermore, strongly territorial animals given to a freely roaming, solitary existence are not likely to acclimate very well to being confined in close quarters with others. And animals adapted to specialized diets are of very limited usefulness as domesticates. Goats, on the other hand, feed on a broad range of plant material, and sheep thrive in rolling grasslands. Neither species is naturally endowed with a capacity for rapid flight or a territorial disposition. They forage over relatively small home ranges in large, highly structured groups led by a single dominant male. Hence they are predisposed to tolerate confinement in close proximity to other members of their species and to behave in a submissive manner. Given these factors in addition to their already being a source of food when still in the wild state, it is no surprise that goats and sheep were among the first animals to be domesticated anywhere in the world.

Just as domesticated plants exhibit differences from their wild ancestors, so do domesticated animals. A very general feature of domesticated animals is a reduction in overall body size. A number of factors have been cited to account for this development, among them confinement in crowded, unhealthy conditions, poorer diet (especially for growing young animals and their mothers, both before birth and thereafter), and herd management practices (e.g., selection of animals for slaughter). Changes in the horns of both goats and sheep are also a feature of domestication. Most remarkably, domesticated female sheep (except in rare instances) have no horns at all, while their wild counterparts seldom lack them. Domesticated pigs have much smaller teeth than their wild ancestors. Once these patterns have been established, they can be used to assign dates to transitions (at given sites) from hunting particular species of animals to managing domesticated varieties of them. Unfortunately, not much is known concerning the specific steps that were taken to domesticate these animals. Since it is a common practice among hunter-gatherers to capture the young of wild animals and to raise them to adulthood, it is not unreasonable to suppose that such a procedure played a role in the domestication of farm animals. Given such uncertainties, the timelines for the developments discussed in this chapter remain subject to debate and revision.

A third important feature of transitions to the agricultural way of life is that they involve the confluence (from a variety of sources) of complementary sets of knowledge, skills, and resources. This point is well illustrated by the appearance of farming communities in the Fertile Crescent some 8,000 years ago, the earliest known anywhere. This is also a useful case study because the advent of agriculture in the Fertile Crescent has been so extensively studied.

B. First Case Study: The Fertile Crescent

Traditionally, the transition to the agricultural way of life in the Fertile Crescent (see Figure 5-1 below) has been called the "**Neolithic Revolution**" because it was accompanied by a new method of fashioning stone tools: grinding and polishing (in contrast to flaking). As discussed in Chapter IV, such tools are much more effective in woodworking than are flaked tools. Although the repertoire of these farmers included quite a variety of crops and a number of domesticated animals, seven of them formed the central core of this new agricultural economy: emmer wheat, einkorn wheat, barley, goats, sheep, pigs, and cattle. The places and times

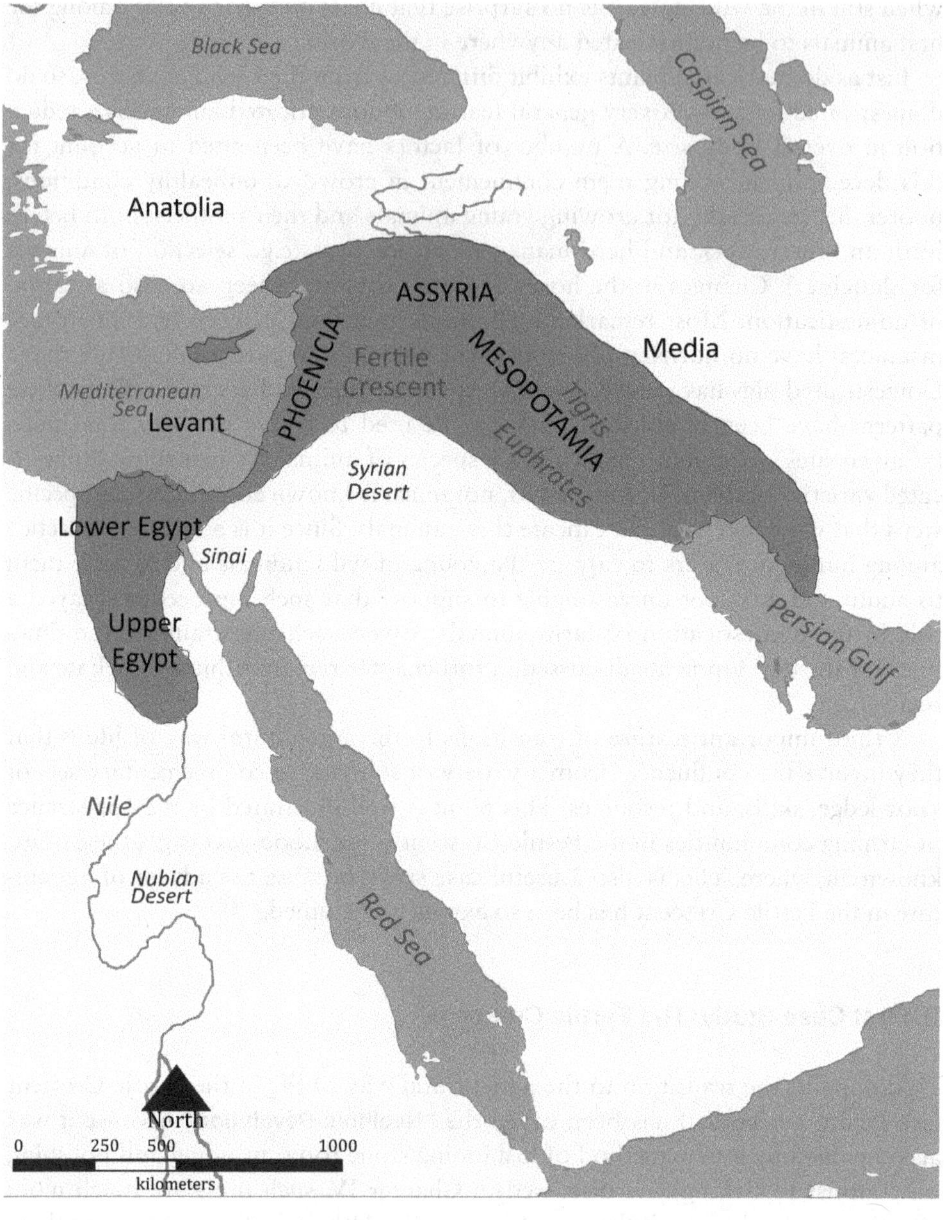

Figure 5-1. The Fertile Crescent. Image credit: Nafsadh. CC-BY-SA 4.0

of domestication for these seven are spread across Southwest Asia from the Levant at the eastern end of the Mediterranean Sea to the Zagros Mountains along the eastern edge of the Fertile Crescent and across close to a thousand years of time.

The first of this group to be domesticated were the grain crops—all three of them—between 10,300 and 9,600 years ago in the narrow strip of land in the southern Levant mentioned earlier. This does not mean, however, that the domestication of these three cereals was a single process or was brought about by the same people. Rather, the conditions at the time and place (including the characteristics of the three plants) were simply ripe for just such innovations. The evidence indicates that each of the small, sedentary communities in the area at the time developed its own way of sustaining itself by a combination of hunting, gathering, harvesting wild grains, and/or cultivating one or more of the three cereals. Cultivation of the domesticated grains gradually spread across the Fertile Crescent, with farming villages scattered from the Levant to the Zagros Mountains by some 9,000 years ago. But not all communities took up the planting of crops, by and large only those living where conditions made it advantageous. Most importantly, that means conditions compatible with a sedentary or semi-sedentary way of living—since, as noted in the Introduction, benefiting from planting seeds requires that one be around to harvest and store the crop. Also included would be the availability of other significant sources of food, given that a vital part of the agricultural juggernaut discussed in the Introduction—domesticated animals—was still missing.

Goats were the first of the animal species to be domesticated. According to the archaeological evidence, this took place some 9,000 years ago in the Zagros Mountains along the eastern edge of the Fertile Crescent. Thus, in Southwest Asia, the domestication of plants preceded that of animals by something like a thousand years. It has been argued that there are reasons for this order. Certainly, the ability to cultivate crops bears upon being able to take responsibility for domesticated animals' getting enough food. The cultivation of cereals had reached the eastern part of the Fertile Crescent by 9,000 years ago, but it is not known (although it seems likely) that those responsible for the domestication of goats were already planting crops. In any case, goat herding rapidly spread west into areas where the cultivation of cereals was well established. Bringing a source of animal protein (meat) under human management was a major step toward making agriculture a stand-alone system for providing for human sustenance. Between 8,700 and 8,200 years ago, sheep were domesticated in the northern (central) part of the Fertile Crescent (roughly, Syria and southeastern Turkey). During this same period, pigs were domesticated in an area overlapping the western part of the same region and extending farther south in the Levant.

Given the importance of cattle in modern agriculture, it will come as a surprise to many that cattle, the last of the four animal species listed to be domesticated, played only a secondary role in the emergence of the new agricultural economy. It is very likely that their huge size was a factor. The wild ancestor of cattle (the aurochs, now long extinct) was up to six feet high at the shoulder and weighed as much as 1,500 pounds. There is compelling archaeological evidence that, in the case of cattle, a marked reduction in size accompanied domestication. In view of the subsequent economic importance of cattle, it is reasonable to suppose that various selection pressures operating during the domestication process resulted in a smaller, more manageable animal. This occurred in the western part of the Fertile Crescent and Anatolia between 8,000 and 7,000 years ago, bringing to completion the set of domesticates that formed the economic foundation of the burgeoning growth of the human population that was to follow. The domestication of cattle also opened the door to another innovation that was to have profound effects on human affairs: the harnessing of **animal power** to carry out more quickly or more efficiently a great variety of operations in agriculture, transportation, warfare, etc. The key role played by animals in such endeavors continued well into the twentieth century in industrialized societies and continues to this day in certain places and for certain purposes.

The confluence of knowledge, skill, and resources required for the growth of agriculture raises the question of how such knowledge was passed from generation to generation. The case of the Bantu expansion, which occurred thousands of years after the agricultural transition in the Fertile Crescent, is presented later in this chapter to demonstrate the role that language might play in such information transmission (though it is unclear to what extent language was involved in the agricultural transition in the Fertile Crescent). As in Chapter III, though, it should be noted that the cognitive capacities of such people would have included trial-and-error learning that could have been applied to learning by watching the outcomes of continued crop planting or animal breeding. One need only notice that from generation-to-generation, certain characteristics are preserved or more likely to occur to attempt planting more of the crop with the desired characteristic. Those early planters need not have realized how different, and how improved, the end result might be over centuries of selective breeding to sustain the practice of attempting to recreate the crop or animal with the best characteristics. For early people still struggling to survive and supplementing their agricultural gains with hunting and gathering, even slight preservation of a desirable trait in an animal or crop over a few generations would have been noticeable.

C. Additional Basic Features: Women's Roles, Larger Families, and Expanding Populations

A fourth point that needs to be made regarding the transition to agriculture is that the practice of planting seeds to augment what nature supplies was almost certainly initiated by women. The basis for this conclusion is simply that, among foragers, women (rather than men) typically are the ones who gather vegetable foods and therefore have the knowledge of plants, their life cycles, and environmental relationships requisite to success in such an undertaking. Furthermore, since women (not men) had always processed most, if not all, of the food items they brought in, and since they were the primary providers of care to infants and very young children, it was almost certainly the women of the early agricultural villages who initiated the practice of feeding the softer, more digestible foods then available to children at an earlier age than is possible among foragers. This development, together with sedentism and other aspects of the agricultural way of life, fueled a dramatic increase in the growth rate of the human population. It came about primarily as the result of the removal of some of the constraints that had limited the number of children a woman was likely to have.

Lacking other suitable foods for very young children, mothers in hunter-gatherer societies must nurse their offspring for longer periods (up to four years) and more intensely than mothers in farming communities, who (as noted in the Introduction) need to nurse their babies less frequently as they consume more and more food from other sources. Since nursing tends to suppress ovulation, women in foraging societies are less likely to become pregnant as soon after giving birth as are women in agricultural societies. Consequently, their children are more widely spaced and fewer in total number. It should be noted that this lower birthrate is an advantage in hunter-gatherer societies. A woman carrying an infant and accompanied by a child about four years old can forage for provisions much more effectively than a woman carrying an infant while herding three or four additional young children, one of whom might be as young as two years old. The severity of the challenge posed to foragers by too many very young children is evidenced by such things as taboos against sexual contact between spouses for an extended period following a birth, interpretations of the arrival of twins as a sinister omen of impending misfortune, and even practices of infanticide. In contrast, children are a significant economic asset among agricultural people. They can weed the garden, herd sheep, stand watch over orchard, garden, and field, driving away avian and four-footed raiders (summoning help if needed), etc. Children playing a significant role in the economic activities of their families is another feature of human life that was ushered in by agriculture; it persisted until the Industrial Revolution.

The transition to the agricultural way of life, involving as it did not only a permanent residence but also a much larger number of children, had major implications for the lives of women. Having a permanent residence did relieve women of the necessity of carrying their very young children with them wherever they went, but this relief came at a price: confinement to the general vicinity of that residence. And the larger number of children women were having meant that, more than ever, women's daily activities revolved around caring for children. As anthropologist Judith Brown has pointed out, the other tasks for which women have assumed responsibility were and had to be compatible with looking after children. Such tasks must not require one's undivided attention, must not pose a danger to children, and must lend themselves to interruption and resumption. The agricultural way of life added another requirement: tasks must lend themselves to being performed close to home. Not surprisingly, the list of tasks traditionally performed by women until relatively recently fit these criteria: spinning, weaving, sewing, food processing (including cooking), and gardening. Of course, women helped out with other tasks (for example, harvesting) as time and circumstance permitted or demanded. Perhaps it should be pointed out at this juncture that the incompatibility of a task with taking care of young children has nothing to do with whether or not women are capable of performing that task.

Inevitable results of agricultural peoples' having large families are burgeoning populations and the need for more land to cultivate. Farming populations grow and spread, occupying larger and larger areas and either absorbing or displacing smaller and more widely dispersed groups of hunter-gatherers encountered in the process. And, of course, the farmers carry their language and culture with them as they move into new territories. Because of their inferior numbers, foragers into whose territory farmers move have little choice but to retreat into areas less suitable for agriculture or to take up farming themselves, in which case they are likely to be absorbed into the more numerous cultural group.

Those learning the agricultural practices of the newcomers can be expected to adopt other cultural elements closely related to these practices as well, language included. Continued over extended periods of time, such processes would appear to account for one of the most striking features of the human cultural scene: closely related languages ("language families") spoken across vast areas. The suggested explanation is that these languages are similar in vocabulary and grammatical structure because they are "daughter languages," languages that diverged from the same ancestral tongue as the descendants of a growing population of farmers became more and more dispersed and separated into subgroups with little contact, the focus of the second case study of this chapter.

But the overall process is a bit more complicated than it might at first appear.

As noted above, fully developed agricultural systems do not spring into existence all at once but come into being only after a number of plants and animals have been domesticated such that, taken together, they can supply all or most of people's nutritional needs under a range of conditions. And certainly the members of such assemblages can be expected to have originated at different times and places and among different peoples and to have been brought together over an extended period of time. Thus an important dimension of the development of agriculture is the adoption of additional farming and/or herding methods by people already engaged in such activities, as exemplified by the adoption of goat-herding by (part-time) grain-farmers in Southwest Asia, as described above, and as powerfully demonstrated in the second case study of this chapter.

When two expanding farming populations confront each other, it is not necessarily the case that either of them will be absorbed or displaced by the other. (Growth simply may take place in other directions, a development also seen in the second case study.) What is likely as a result of such encounters (unless the two groups have the same technological resources) is that one or both will add to its repertoire of practical arts by "borrowing" from the other. This is likely to be accompanied by the adoption of other cultural elements (such as language) closely associated with a borrowed technique or device. Non-literate cultures have relatively small vocabularies highly correlated with their practical activities and concerns (as discussed in Chapter VI). Consequently, those adopting a technology from another culture usually do not have readily available words needed to communicate effectively regarding it. Accordingly, the common practice is to use expressions borrowed from the donor society (a major theme of the second case study).

Another characteristic of such technological transfers is that they often include cultural elements that would be judged by contemporary standards to be inessential to the practical art in question. The explanation for this feature of technological "borrowing" seems to be that those engaging in it are in no position to distinguish between actions or conditions that are essential to a technology (as practiced by the donor culture) from those that are not, such as rituals or ceremonies. An example can be found in the story of Cain and Abel. Scholars believe that the ancestors of the Ancient Israelites were wandering herders who took up the practice of agriculture only after settling in the Promised Land, where they learned to farm from those already living there. Unfortunately, the procedures for growing grain that they adopted included religious observances which the Israelites had no way of distinguishing from such matters as selecting a suitable plot of ground, preparing a seedbed, planting at the appropriate time, etc. Consequently, those Israelites who took up growing crops in this manner were in violation of one of the basic tenets of their own religion, which forbade honoring the claims of any other religion in any way. Genesis gives no explanation for God's rejecting Cain's sacrificial offering from his grain crop while accepting Abel's offering from his flock of sheep. The interpretation favored by scholars is that while grain, as such, would not necessarily have been unacceptable as an offering, grain which had been produced in a manner involving paying homage to another god would have been.

The absence of a distinction between technological practices and ceremonies or rituals (whether religious or otherwise) is not always problematic. Indeed, the observance of rituals in the process of carrying out some technological procedures can be essential to ensuring that the sequence of essential operations is carried out correctly and precisely. This is especially true in the case of practical arts that involve complex steps such as making steel prior to the modern era; the tradition of crafting swords by Japanese artisans using symbolic color cues demonstrates this well. Steel is an alloy, a mixture of iron and carbon—a very tiny percentage of carbon. Producing it involves precise control of the carbon content and heat treatment. The utility of steel derives from its being hard and flexible rather than hard and brittle like cast iron, making it ideal for weapons such as swords.

A material that combines hardness and flexibility is possible if it consists of layers. Traditional ways of achieving this include cutting a billet of metal into two pieces, stacking one piece on top of the other, and welding them together by heating and pounding. Since the properties of each layer are not identical to those of the next (because they do not have identical histories of being heated and pounded), the bond between them is not as rigid as it otherwise would be. In traditional methods, these steps are repeated over and over, each time doubling the number of layers. This may be done as many as fifteen times, as seen in the forging of steel swords for the fearsome samurai of Japan (see Figure 5-2 below). The final step in making a steel sword the traditional way is known as **quenching**, plunging the blade into water after heating it. The rapid cooling thus produced hardens the metal. In order for this procedure to achieve the desired result, the blade must be heated to an established level, which (in the absence of instruments for measuring temperature) apprentices learn to judge on the basis of color cues. Paralleling a pattern seen in the earlier stages of sword making, the judgments of a Japanese artisan regarding the temperature of a blade being heated are informed by a long-standing tradition into which he has been initiated and which includes (as a color cue) the imagery and symbolism of the morning sun. Other rituals rich in symbolism, but now lost to time, may likewise have guided the transmission of information related to technology, especially agricultural

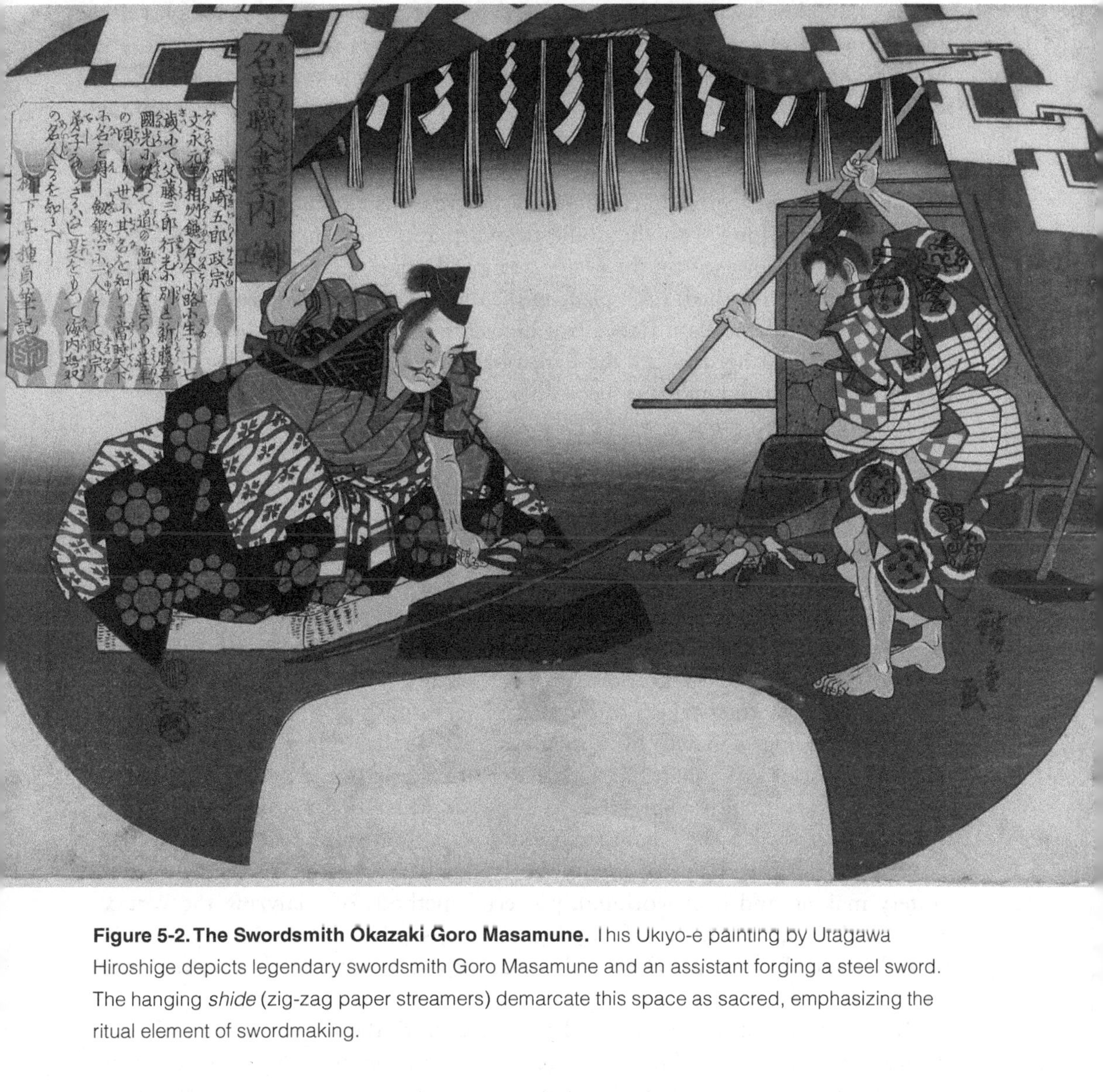

Figure 5-2. The Swordsmith Okazaki Goro Masamune. This Ukiyo-e painting by Utagawa Hiroshige depicts legendary swordsmith Goro Masamune and an assistant forging a steel sword. The hanging *shide* (zig-zag paper streamers) demarcate this space as sacred, emphasizing the ritual element of swordmaking.

technology, in earlier cases. Today, for example, rituals still guide agricultural practices in East Java. There, long-established, traditional rituals such as *Sedekah Bumi* (earth offering) transmit ecological know-how concerning factors like planting timing and soil preservation.

The examples discussed above amply demonstrate that the absence of a distinction between technology and other cultural elements can—but need not—be an obstacle to receiving whatever benefits a given technology has to offer. The following case study chronicles the step-by-step process by which Bantu-speaking people repeatedly added to their farming and herding repertoire by "borrowing" domesticated plants and animals as well as practices from groups they encountered as they spread across Africa. For reasons explained above, it is not surprising that, along with the domesticated plants and animals, they adopted the names for them used by the donor groups. As one might expect, this "Bantu expansion" (as it is called) took place in the direction of the foragers living to the east and south of the Bantu ancestral homeland rather than in the direction of other farmers and herders. As the Bantu became more and more dispersed and out of contact, their languages became increasingly divergent though similar in vocabulary and grammar.

D. Second Case Study: The Bantu Expansion

Bantu languages, which are spoken across a vast swath of central and southern Africa, are believed to have diverged from an ancestral tongue once used by people who inhabited what is now southeastern Nigeria and western Cameroon. It appears that an out-migration from this area got underway between 2000 and 1500 BCE, with speakers of Bantu languages pushing east and south along routes still being debated by specialists. Eventually, Bantu-speaking people occupied most of sub-equatorial Africa except for deserts and other areas unsuitable for their agricultural practices. This great watershed in African history is often referred to as the "Bantu expansion." Although physical (archaeological) evidence of the chains of events involved is limited (primarily to vestiges of pottery making and iron working), powerful methods of analyzing the syntax, vocabularies, and phonetics of existing languages (often referred to as "linguistic archaeology") have provided truly amazing vistas on the prehistory of Africa (and other places as well).

Once a family of languages has been recognized, the next step is to identify the features that the daughter languages all have in common. Absent evidence to the contrary, these are assumed to have been present in the ancestral language. Expressions that are present in only some of the daughter languages are assumed to have been acquired after the speakers of those languages became geographically separated from the others. And similar reasoning can be applied to each subset of the daughter languages with members that all share one or more of the features not present in the ancestral language. If some (but not all) members of such a subset share additional features not present in the ancestral language, then those features may be assumed to have been acquired after the speakers of those languages became geographically separated from the speakers of other languages in the subset. And so on. For example, the word for (African) yam in all Bantu languages are cognates, but the words for chicken are not, indicating that the former (but not the latter) had a place in the ancestral Bantu language and that the different words for chicken in the various Bantu languages were acquired after the relevant populations had become geographically separated. One of the loanwords most frequently found in Bantu languages is a word for sorghum. Somewhat less widely present is a loanword for iron. The implication is that the word for iron was acquired sometime after the word for sorghum and after some Bantu populations had become geographically separated from the others.

Repeated applications of this kind of reasoning have been used to construct a chronology of the development of Bantu vocabularies and syntax. Most importantly, for present purposes, the result is also a chronology of the Bantus' cultural development—technology included. The assumption is, of course, that if, by a certain point, a people have a word or expression for a thing or process, then (in the absence evidence to the contrary) they can be assumed to have been acquainted with it; and if, prior to a given time, they had no word or expression for something, they probably were unacquainted with it. Linguists working to recapture the development of Bantu languages have had the distinct advantage that the bulk of the ancient languages from which Bantu loanwords were drawn still survive. If the language from which a loanword comes is known, then contact with the corresponding culture can be inferred—and so can the presence in that culture of the object or practice to which it refers. Furthermore, if the item in question is a technology subsequently practiced by the culture using the loanword, it may reasonably be concluded (again, in the absence of evidence to the contrary) that it was through such contacts that these people learned of it. Using the method outlined, archaeo-linguists have been able to identify a series of Bantu contacts with other cultures through which they acquired key elements of the technological repertoire that enabled them to overspread so much of sub-equatorial Africa.

While still living at their original location, the Bantu had mastered the art of growing yams and that of making polished stone tools, most notably axes, which would have been the key to clearing plots for growing these plants in the forests where they lived. (As discussed in Chapter IV, stone tools made by grinding and polishing are much more effective than flaked ones for cutting wood.) Evidence of such tool making has been found in the archaeological record. At the time of their initial migrations, the Bantu—and the people they would be encountering and learning from—would have been sustaining themselves by one combination or another of hunting, fishing, gathering, and practicing whatever agricultural skills they had mastered. As they migrated across Africa, the Bantu acquired new technologies from the people they met and, as one might expect, adopted the relevant vocabularies of those from whom they acquired technologies. As noted above, the Bantu word for sorghum is such a loanword. It was "borrowed" from the Central Sudanians whom the Bantu encountered as they approached Africa's (Western) Rift Valley from the west. The core of the material culture of those people was grain farming, particularly sorghum and pearl millet. The Bantu probably learned to grow both crops from the Central Sudanians (although the linguistic evidence is not as clear in the case of pearl millet).

A little farther to the east, the Bantu came into contact with the Eastern Sahelians, from whom they adopted even more important practices and devices, as the Eastern Sahelian loanwords in Bantu languages testify. Among others, these include words or expressions for livestock pen, fence, dewlap, fresh milk, flour, wooden bowl, and iron (points). It follows that, as a result of their contact with the Eastern Sahelians, the Bantu probably took up the practice of animal husbandry (most likely involving goats), constructed pens for their livestock, and obtained milk from them. Furthermore, they probably began using **iron** spear points and arrowheads at this juncture. Somewhat later, as Bantu populations spread to the south and fanned out, Eastern Sahelian loanwords for sheep could be found in some of their vocabularies. By roughly 500 BCE, **cattle herding** had become an established part of the local economies of at least some Bantu communities living in the area to the southwest of Lake Victoria, as is attested by the Eastern Sahelian and Southern Cushite loanwords for cow or cattle in their vocabularies. The contact that the Bantu made with the Southern Cushites (then present in the area to the southeast of Lake Victoria) is especially important because of the advanced state of their pastoral practices. Evidence of the significance of this interaction is afforded by the presence in Bantu languages of specialized Cushite loanwords for such things as young cattle (female or male) that have not yet reproduced. The Bantu and the other groups discussed also benefited from a number of additional, but less significant, domesticated plants: black-eyed peas, Voandzeia groundnuts, sesame, gourds, cotton, etc.

The above is by no means intended to be a complete account of the crafts, practices, and resources that supported the Bantu expansion, but it does identify the primary technological engines that powered it. With the technologies described at their command, the Bantu would be expected to have grown in numbers and to have taken advantage of opportunities to move into areas suitable for farming but inhabited only by scattered bands of hunter-gatherers. And that is just what they did. Prior to the arrival of the Bantu, neither the inhabitants of the Congo Basin nor those of the whole of sub-equatorial Africa (with the exception of the Lake Victoria area) practiced agriculture. Indeed, no plant or animal native to sub-equatorial Africa had been domesticated by the time the Europeans arrived in the seventeenth century. The crops and livestock vital to the Bantu way of life were all brought in from outside the region. Thus, the stage was set for a classic example of a highly developed agricultural society overwhelming scattered communities of hunter-gatherers. Long before the Europeans had begun to put their mark on sub-Saharan Africa, Bantu-speaking farmers and herders had occupied a vast swath of the continent from the Congo Basin almost to its southern tip.

Along the way, the Bantu arrived on the east coast of sub-equatorial Africa, which set the stage for their acquisition of some important new crops and a domesticated fowl. This came about in conjunction with one of the most remarkable episodes in human history: the colonization, early in the first millennium CE, of the island of Madagascar (previously uninhabited). Surprisingly, the colonists were Indonesians, not Africans! About the same time, or just before, some of the Indonesians came ashore on the east coast of Africa, bringing with them bananas, Asian yams, taro, sugarcane, and chickens. These became established among local Bantu farmers and spread inland from one community to the next and were carried into new territories as well.

At this juncture, some 1,000 years after they began their out-migration, the Bantu collectively were in possession of an extensive repertoire of domesticated animals, wet-climate crops (from their ancient homeland and from the Indonesians) and dry-climate crops (obtained from the Central Saharians, the Eastern Sahelians, the Cushites, etc.). This does not mean that an individual Bantu speaker or a group of them could expect to enjoy a diet encompassing all of these crops. Rather, those living in a given locale had a diet featuring crops appropriate to the climate of that area. And, of course, they all helped themselves to nature's bounty by fishing, hunting, etc. Overall, they enjoyed rich and nutritious diets. A key factor underlying the spread of the Bantu speakers across so much of Africa was their readiness to adopt new domesticates and practices from other groups. Thus, they were able not only to survive but even prospered under a variety of climatic conditions. It would be difficult to exaggerate the importance of this feature, both with respect to the Bantu and to the foragers with whom they came into contact.

The fate of the original inhabitants of the territories into which the Bantu expanded followed predictable patterns. Some of them were absorbed into the dominant culture; some remained in or retreated into areas unsuited to the agricultural practices of the newcomers; some adopted one or more of the survival strategies of the Bantu while maintaining their own separate cultural identity; and some simply did not survive. By the time the Europeans arrived in the fifteenth century, the original inhabitants of the central African rainforests, the Pygmies, lived in scattered pockets in the primeval forest left undisturbed by the Bantu as they navigated the rivers in their dugout canoes and cleared plots in which to grow their crops. Although the Pygmies did not take up the farming practices of the Bantus, they did trade with them and worked for them. As a result of this interaction and of the fragmentation of the Pygmy population into isolated pockets, each group of the diminutive hunter-gatherers learned the particular Bantu language spoken by their farmer neighbors and, over time, abandoned their own language, retaining only a few words and distinctive pronunciations. Given that language is both an important part of any culture and a repository of it, the loss sustained by the Pygmies must have been substantial.

Another distinct group of hunter-gatherers who were engulfed by the Bantu farmers and herders are the Khoisan, who at one time, it seems, roamed across the African continent from just to the south of Lake Victoria to its southern tip. But by the time Europeans arrived, the Khoisan could be found only in the far south and southwestern part of Africa (roughly below the northern border of present-day Namibia), much of it unfit for cultivation and the remainder unsuited to the agricultural practices of the Bantu who had advanced down the east coast. (The principal grain crops of the Bantu were sorghum and pearl millet, which are adapted to a seasonal pattern of summer rains, whereas the southern coastal area of Africa has a "Mediterranean" climate: winter rains and summer drought.) The Khoisan can be distinguished from other sub-Saharan Africans by their lighter complexion, their more tightly coiled hair, and, among other things, the use of click sounds in their speech. At one time two distinct divisions of these people were recognized: (i) the San, who are relatively small in stature, and (ii) the taller Khoikhoi. Although neither of the Khoisan groups had the use of iron (a highly developed technology among the Bantu), the Khoikhoi did take up the practice of herding sheep and cattle. The San remained hunter-gatherers, but they also worked for the Khoikhoi pastoralists. The arrival of the Europeans with their even more powerful technologies (including firearms and more advanced farming tools and methods), crops suited to South Africa's "Mediterranean" climate, and new breeds of livestock carrying new diseases with them posed new challenges for the Khoisan. Since so many of the Khoikhoi were killed, died of the new diseases, or were absorbed into the emerging complex South African culture, they have effectively ceased to exist as a distinct group. The San were able to maintain their traditional hunter-gatherer way of life until late in the twentieth century in the desert shrub areas unfit for agriculture.

E. Retrospect and Prospect

Both of the case studies included in this chapter provide an illustration of a fundamental feature of the genesis of major technological systems. Rather than being created deliberately to perform the function(s) that they or their extensions or adaptations eventually came to serve, such systems come about through a series of additions or steps, each with characteristics or aspects heavily dependent upon local conditions at the time. This pattern shows itself in the invention of writing, a topic to be addressed in the following chapter (VI), as well as in the development of other major technological systems discussed in subsequent chapters. These processes are by no means linear and accumulative. They proceed in fits and starts and may have features (along the way) that are not destined to have a place in what eventually emerges. Although it cannot be inferred from linguistic evidence alone, the spread of Bantu languages into eastern and southern Africa (carried along by the rising tide of farming and herding) was by no means smooth and consistent in direction, especially on a small scale.

The agricultural way of life involves an enforced sedentism that bestows upon its practitioners the opportunity for enhanced technological development like that enjoyed by the Magdalenians and the Native Americans of the Pacific Northwest (discussed in Chapter IV). Having a more or less permanent place of residence enables people to use a greater variety of instruments and materials and provides them with more opportunities to invent new devices or discover new uses for old ones (as discussed in Chapter IV; see also Chapter II, Section A). In this way, agriculture set in motion the snowballing accumulation of technological capabilities so much in evidence in today's world. Additional effects include larger families, major changes in the lives of women, the need for more land, and pressure on neighboring populations of foragers from the out-migrating agriculturalists. Over time, language families like that of the Bantu result.

Another, the Indo-European language family, was the first to be identified. Its members are believed to be descendants of the same ancestral language. This conclusion resulted from the recognition of similarities between the vocabulary and grammar of (i) Sanskrit, the ancient Indic language of the Hindu sacred writings, and (ii) those of classical Greek and Latin. Since the existence of an ancestral language was postulated in the 1780s, a great deal of progress has been made in identifying the members of the Indo-European language family, which includes most of the languages of Europe as well as Persian (or Farsi) and the Indic languages of southern Asia, and in reconstructing the vocabulary, syntax, and even the phonology of the ancestral language. Knowledge of the language of the Indo-Europeans has, of course, provided much information about their culture, including their technological resources. Non-literate cultures have relatively small vocabularies that are highly correlated with their practical activities and concerns (as discussed in Chapter VI). But despite concerted efforts to pin down who they were and where they lived, scientists and scholars remain deeply divided regarding the answers to these questions. Hypothesized homelands of the people who spoke the Indo-European ancestral language range from central Europe to the Asian steppes (north and east of the Caspian Sea) and from the Volga River basin to Asia Minor.

It seems natural for human beings to ask, "Who am I?" The ancient Delphic oracle even made it a duty: "*Know thyself.*" Clearly, more is involved here than knowing the answer to the question frequently posed by adoptees, orphans, etc., "Who are (were) my biological parents?" or to questions of the type repeatedly posed by people seeking names and dates to enter into a family tree. Since so much of people's identity is bound up with the language (and the culture) of their formative years, all who recognize the sagacity of the Delphic injunction should seek to identify the developments that have contributed important elements of that language and culture. Such questions assume added urgency when people find themselves caught up in swirling linguistic and cultural crosscurrents.

Language is both an important part of culture and a repository of it. Languages involve categorizations, patterns of thinking, assumptions about people and the world, value commitments, etc. Of paramount importance because they are foundational to thought and communication are ways of categorizing things (viewing certain ones as instances of the same *type* of entity or occurrence) and words or expressions for these groupings. Also fundamental to these processes is the drawing of distinctions, the assertion of the need (for certain purposes) to distinguish between two (or more) types of things. For example, it is important to understand that, despite superficial resemblances, some mushrooms are edible while others are extremely poisonous. An example relevant to current public discussion is the difference between the question of whether something is ethical and the question of whether it is legal. Among the factors that have made major contributions to the makeup of languages, the invention of writing stands out. This topic is the subject of the next chapter.

Sources / Further Reading

Barber, Elizabeth Wayland. *Women's Work: The First 20,000 Years*. W.W. Norton & Company, 1994.

Bellwood, Peter, and Colin Renfrew, editors. *Examining the Farming/Language Dispersal Hypothesis*. Cambridge: McDonald Institute for Archaeological Research, 2002.

Bellwood, Peter. *First Farmers: The Origins of Agricultural Societies*. Blackwell Publishing, 2005.

Bhanoo, Sindya N. "Farming Had an Earlier Start, a Study Says." *New York Times*, 28 July 2015, p. D4.

Bronowski, Jacob. *The Ascent of Man*. Little, Brown and Company, 1973.

Brown, Judith K. "A Note on the Division of Labor by Sex." *American Anthropologist*, vol. 72, no. 5, October 1970, pp. 1073–78.

Cowan, C. Wesley, and Patty Jo Watson, editors. *The Origins of Agriculture: An International Perspective*. Smithsonian Press, 1992.

Danugroho, Agus, Rustinsyah Rustinsyah, Muhammad Adib, Iwin Ardyawin, Usman Idris, and Muhammad Alhada Fuadilah Habib. "Ritual, Ecology, and Climate Change: An Ethnoecological Study of Agricultural Communities in East Java." *Social Sciences & Humanities Open* 12, 2025, 102061.

Diamond, Jared, and Peter Bellwood. "Farmers and Their Languages: The First Expansions." *Science*, vol. 300, no. 5619, 25 April 2003, pp. 597–603.

Diamond, Jared. *Guns, Germs and Steel*. W.W. Norton & Company, 1997.

Ehret, Christopher. *An African Classical Age: Eastern and Southern Africa in World History, 1000 B.C. to A.D. 400*. University Press of Virginia, 1998.

Gebauer, Anne Birgitte, and T. Douglas Price, editors. *Transitions to Agriculture in Prehistory*. Prehistory Press, 1992.

Hall, John A., and I.C. Jarvie, editors. *Transitions to Modernity: Essays on Power, Wealth and Belief*. Cambridge University Press, 1992.

Haviland, William A. *Anthropology*, 8th ed. Harcourt Brace College Publishers, 1997.

"The Indo-European Family of Languages." *The American Heritage Dictionary of the English Language*, 5th ed. Houghton Mifflin Harcourt, 2011, inside back cover. Chart.

"Indo-European Roots." *The American Heritage Dictionary of the English Language*, 5th ed. Houghton Mifflin Harcourt, 2011, pp. 2094–2134.

Jones, William B., and A. Warren Matthews. "Toward a Taxonomy of Technology and Religion." *Technology and Religion*, edited by Frederick Ferré. *Research in Philosophy and Technology*, vol. 10, 1990, pp. 3–23.

Kent, Susan, ed. *Farmers as Hunters: The Implications of Sedentism*. Cambridge University Press, 1989.

MacNeish, R.S. *The Origins of Agriculture and Settled Life.* University of Oklahoma Press, 1992.

Mallory, J.P. *In Search of the Indo-Europeans: Language, Archaeology and Myth.* Thames Hudson, 1989.

Renfrew, Colin. *Archaeology and Language: The Puzzle of Indo-European Origins.* Cambridge University Press, 1987.

——. "World Languages and Human Dispersals: A Minimalist View." *Transitions to Modernity: Essays on Power, Wealth and Belief*, edited by John A. Hall and I.C. Jarvie. Cambridge University Press, 1992, pp. 11–68.

——. "World Linguistic Diversity." *Scientific American*, vol. 270, no.1, January 1994, pp. 116–23.

Smith, Bruce D. *The Emergence of Agriculture.* Scientific American Library, 1994.

Solbrig, Otto T., and Dorothy J. Solbrig. *So Shall You Reap: Farming and Crops in Human Affairs.* Island Press, 1994.

Watkins, Calvert. "Indo-European and the Indo-Europeans." *The American Heritage Dictionary of the English Language*, 5th ed. Houghton Mifflin Harcourt, 2011, pp. 2081–89.

Zohary, D., and M. Hopf. *Domestication of Plants in the Old World.* 2nd ed. Clarendon Press, 1993.

Chapter VI:
Record Keeping, Counting, Writing, and the Literacy Revolution

WITH THE POSSIBLE EXCEPTION of agriculture, no other technological innovation has come close to having as profound an effect upon human life as has making **physical records of information**, the technology that (eventually) gave rise to writing and mathematics as we know them. In this case, above all others, the enormous difficulty that people encounter in imagining being without a technology stands stubbornly in the way of their appreciating just what an achievement these inventions were. Most significant, in this case, is the near impossibility of thinking like someone who is unacquainted with writing and counting and their uses. The magnitude of this challenge is illustrated in a powerful way by the absence, prior to the 1930s, of any real understanding of oral cultures and non-literate patterns of thought—not only by the general public but on the part of scientists and scholars. Up to that point, scientists and other thinkers (themselves literate!) had just assumed that only differences in factual information, belief systems, ways of classifying things, and the like distinguish literate people from those who do not have the use of writing. As long as this view prevailed, it was impossible for literate persons to recognize the profound changes that had come about following the invention of writing. There had to be, in effect, a rediscovery of non-literate thought. Only when this had been achieved was it possible to grasp what a profound transformation the "**literacy revolution**" really was.

Also indicative of the difficulty that (numerate) people have in grasping the significance of the invention of physical records is the tendency to regard **counting** as a "natural" activity, not unlike talking. Such a view is suggested by the use of the expression "natural numbers" to designate the positive integers and by the pronouncement, attributed to the mathematician Leopold Kronecker (1823–91), that "God made natural numbers; all else is the work of man." To those who have struggled in math class, the positive integers probably do seem "natural" in comparison to common fractions, decimal fractions, irrational numbers, transcendental numbers, negative numbers, imaginary numbers, complex numbers, etc. However, contrary to what one might expect, all people and cultures do not have, as a kind of natural attribute, a system of numerical concepts and a way of counting like those familiar to every schoolchild. Rather, these appear to emerge only insofar as a society has a need for them and they serve a useful purpose. Since non-literate, hunter-gatherers like the Khoisan of southern Africa have very little need for such things, they typically develop only a very rudimentary set of quantitative terms and concepts, corresponding to "one," "two," and "many." Attempts by anthropologists to elicit responses indicating an ability to distinguish between, say, fifty and a thousand are met with incomprehension.

A second barrier to understanding the emergence of writing and counting is the presumption, on the part of those familiar with them and their and uses, that it was for just such purposes that these technologies were devised. However, just as was the case with the development of agriculture, writing and number systems did not result from efforts to obtain the benefits such systems are now known to make possible. They were not produced by people setting out to devise ways to record human speech or to create a single system that could be used to tally pitchers of beer as well as fresh fish. Rather, each society has developed different ways of keeping up with different kinds of things and different ways of making sure that important quantitative conditions are met (e.g., dowries or bride prices) according to their needs. As societies become more complex, they add to and/or refine these systems, usually in a step-by-step fashion, in many cases realizing possibilities not likely to have been anticipated prior to the previous step. And while these developments may seem to the modern reader to have been successive steps toward an obvious goal, they actually may have occurred quite independently of each other.

Over time and around the world, human beings have devised a great variety of systems for recording and transmitting information. Just surveying the major record keeping and writing systems that the world has known—leaving aside questions of their origin and development—would be a formidable undertaking. So, in order to give the reader some sense of the way record-keeping systems develop and the role they play as societies become more complex, this chapter will feature a case study of one such system, that of the ancient Sumerians; others will be discussed only in passing. This example was chosen because the Sumerians' **cuneiform script** appeared earlier than any other fully written language, because Sumerian record keeping has been so thoroughly studied, and because these studies have enjoyed such remarkable success in revealing not only the emergence of cuneiform from an earlier record-keeping system, proto-cuneiform, but also the emergence of the latter from even earlier systems. These sharply focused studies have recently brought to light unanticipated and remarkable insights into the way counting and number systems develop. In the process, it has become increasingly clear that it is not to be assumed that systems for writing numbers are merely a means of recording pre-existing ways of counting and reckoning, just as it is not to be assumed that scripts made it possible to record human speech without thereby facilitating or initiating significant changes in its content and structure. Rather, in both cases increased technological capabilities were accompanied by important changes in human thought and behavior. Yet another reason to study Sumerian record keeping (and later developments based on it) is the continued reliance (right up to the present) on **basic mathematical operations** then used and the presence in contemporary mathematics of elements from these earlier systems. Following the case study, we discuss the nature and significance of the **literacy revolution.**

A word of caution is in order. While it would be surprising indeed if there were no parallels between developments in Ancient Sumer relating to record keeping and writing and those in other urban centers, it must not be assumed that the route taken by the Sumerians is the only possibility. Rather than being content with the present account, it is hoped that the reader will be moved to go beyond it, to seek out other such routes and to explore them.

A. Early Record Keeping in Ancient Sumer

Sumerian record keeping has its deepest roots in a practice probably as old as the human need to keep up with things and yet persists in isolated agricultural villages. Such mundane but vital tasks as making sure that the sheep have all been rounded up at night depends upon being able to count or to keep some kind of tally. One simple and time-honored method, still used by shepherds in Iraq as late as the twentieth century, is to maintain a collection of pebbles (or other small objects), one for each animal in the flock, and to move one of the pebbles from one pile to another to register the presence of an animal. In a similar fashion, it can be known that very definite economic or social obligations have been fulfilled even in the absence of verbal counting as we know it. Such simple procedures exploit the most fundamental of mathematical principles: one-to-one correspondence. Of course, systems of this kind become progressively more unwieldy as the size and the number of the groups being dealt with increases. Not surprisingly, more efficient systems were developed as life became more complex. The story (or stories) of how people came to count verbally as we now do is more complicated than most are likely to imagine, and the details may never be known with any degree of certainty. But there is compelling evidence—spread across southwestern Asia from Turkey to Pakistan—of the widespread use in ancient times of a system for keeping track of things by means of **small clay counters or tokens:** tiny spheres, disks, cylinders, and cones. For a very long time, archaeologists and museum curators had no clue as to the use of these ubiquitous clay objects. Were they children's toys? Gaming pieces? Perhaps those with holes piercing them were beads.

The solution to this puzzle came about through a fortuitous set of circumstances linking the tokens to the earliest Sumerian ideographic and pictographic inscriptions, usually referred to as **proto-cuneiform.** While many details remain unresolved, the chain of connections is as follows. To begin with, one must keep in mind that a bunch of tokens (or pebbles) constitutes a record of a collection of things only if they are kept together and separate from others. At one time groups of tokens probably were kept in a small cloth or leather bags or perhaps wicker baskets, but one would hardly expect anything to remain of such ancient pieces of gear. Artifacts do remain, however, from an intriguing method of keeping sets of tokens together that was employed in the Ancient Near East late in the fourth millennium BCE: enclosing them in **hollow clay spheres or ovoids.** And, as strange as it may seem, these clay envelopes, or *bullae* as they have been called (see Figure 6-1 below), were the key to solving the mystery of the tiny tokens, just as (so long ago) they appear to have been the key to the invention of a record-keeping system from which sprang the world's oldest script.

The special advantage of the clay envelopes seems to have been that **personal seals** could be impressed into their exterior surfaces before they were baked. Then, as long as the envelopes remained unbroken, the seals served to certify their contents. Evidently, the use of personal seals was a well-established practice

Figure 6-1. Clay Envelope and Tokens. This clay envelope (*bulla*) containing tokens and marked with a seal was recovered from an archaeological site in Susa (modern-day Iran) and dated between 4000 and 3100 BCE. Image credit: Marie-Lan Ngyen. CC BY 2.5

in Ancient Sumer when the clay envelopes were first introduced. These seals took the form of engraved cylinders which, when rolled over a surface of wet clay, left the negative impression of the images carved on them. Apparently, their original function was that of a property mark; a small, clay tablet bearing the impression of a person's seal and attached to an object by a string served to identify the owner or sender. Once in use, seals were available for other identification purposes, such as specifying the person responsible for an action or the guarantor of the accuracy of a record, as is now done with signatures. (On some of the envelopes that have been recovered, there are two seals, paralleling the current practice of having two signatures on official documents, one from each party to a transaction.) **Clay envelopes** have been found with their surfaces marked, not only with personal seals, but also with impressions matching, in number and shape, the **tokens contained** within. In one case this correspondence is so exact that one can only conclude that the tokens were pressed into the surface of the wet-clay envelope before being placed inside. This procedure made it possible to know the contents of the clay envelopes without breaking them open, which must have been a significant advantage, as a number of envelopes thus marked have been found.

Since all of the information stored by means of the tokens was also conveyed by the external markings, what seems to have happened next can hardly be surprising: the token-filled clay envelopes were supplanted in a relatively short period of time **by small clay tablets bearing markings** corresponding to the spheres, cones, disks, etc. previously in use. And on many of these tablets appear the impressions of personal seals, presumably serving the same function as before.

By the time tokens were displaced by small clay tablets for record-keeping purposes, the older system appears to have evolved into something much more sophisticated than that employed by non-literate shepherds as described above. Indeed, that may already have been the case even at the earliest date that **tokens** are known to have been present, around 8000 BCE. A large collection of tokens, dating from about that time, recovered at Tepe Isiab in southwestern Iran, consists of fifteen or more distinct types, including spheres, cones, disks, cylinders, ovoids, and even an animal head. Certainly, having different kinds of tokens for different purposes would have made it easier to keep up with things and would have facilitated administrative and commercial activities. Denise Schmandt-Besserat (b. 1933), who has devoted many years to the systematic study of the baked clay tokens of the ancient Middle East, points out that while the basic types of tokens seem to have remained relatively unchanged from the time of their introduction until about 3500 BCE, the next four centuries witnessed a dramatic increase in the number of distinct types of tokens. This increase came about primarily through the proliferation of subtypes or variations of the fifteen or sixteen basic types of tokens mentioned earlier. The distinguishing features of these "complex tokens," as Schmandt-Besserat refers to them, include perforations, indentations, and incised lines. In her view, producing more and more distinct types of tokens was the ancient Sumerians' way of addressing the growing need for ever more powerful record-keeping capabilities in their increasingly complex society. She believes that specific types of tokens came to be representative of definite kinds of animals or products. Furthermore, noting the striking similarity between the shapes of some of the tokens and inscribed signs found on the clay tablets, Schmandt-Besserat argues that the latter are two-dimensional representations of the three-dimensional tokens and that the denotations of these tokens can be inferred from that of the corresponding inscribed signs when the latter can be traced to cuneiform characters of known signification. Should Schmandt-Besserat's account be vindicated, it would surely constitute a powerful synthesis, but it has been subjected to severe criticism, notably by Stephen J. Liebermann and Paul Zimansky. At present there simply is not sufficient evidence to bring the case for her suggestion to closure.

Fortunately, other pieces of the puzzle fit more snugly together. The signs on the earliest clay tablets are very similar to those found on the exterior surfaces of the clay *bullae*. Both are impressions, produced by pressing a solid object (whether it be a token of a round stylus) into the wet clay. In addition to the appearance of the two kinds of artifacts, the circumstances in which some of them were found provide good evidence that the two types of objects had similar functions. Accordingly, since the **impressions on the clay tablets embody** a second fundamental feature of number systems, namely **grouping**, it may reasonably be inferred that the impressions on the clay *bullae*—and thus the clay tokens themselves—do as well. Grouping involves gathering individual units into equal groups and then focusing upon the groups rather the individual units. This procedure greatly facilitates keeping track of large collections or tallies. Anyone who has kept score or recorded votes by means of short vertical lines with every fifth entry a diagonal line drawn across the preceding four has made use of grouping. It is the first step beyond simple tallying, i.e., just adding more and more entries to a record. If uninterrupted, such series of marks (or collections of tokens) quickly blur into "many," but if broken up into equal sub-groups, the human mind and eye is able to focus upon the sub-groups (rather than upon their members) and the question of how many of these there are. Given that the subgroups all have the same number of members, one can know that, for example, one candidate received twenty-eight votes with essentially the same clarity and assurance as one can know that someone is holding up four fingers. Evidently, the Fiji Islanders of yore knew how to use this strategy. A club from the Fiji Islands, preserved in a museum, has a series of notches cut in the handle, presumably to record the hunting or fighting triumphs of the owner. What makes this tally noteworthy is the grouping strategy it embodies: every tenth notch is significantly larger than the preceding nine. By thus grouping the cut marks, the owner of the club was able to have as firm a grasp of their number (fifty or so) as if he had counted them, whether or not he had number words to do so.

Perhaps the most powerful aspect of grouping is its applicability not just to individuals but also to groups, allowing for the formation of **groups of groups**—and, in turn, groups of these groups, and so on. Grouping can be used effectively to make certain that a particular collection of objects is of a required size even in the absence of a (complete) system of number words and even when the collection is quite large. An excellent example, reported by anthropologist Jack Goody (1919–2015), is the traditional method used by the LoDagaa of Northern Ghana to ascertain that the requisite bridewealth payment, which normally adds up to 20,000 cowries (shell money), has been assembled. Faced with a pile of cowries to be counted, a LoDagaa would begin the process by sweeping three of them to one side; then he would gather up two more cowries and add them to the first three, visually confirming (as a check) that a group of five had been formed. After four groups of five cowries have thus been assembled (again a number verifiable by visual inspection), the LoDagaa would gather them together into a pile of twenty cowries. After going through this whole process four more times, he would then collect the five piles of twenty cowries each into one big pile, consisting of one hundred cowries. And so on. Grouping is also fundamental to the number system learned by every schoolchild; in it, units are grouped into tens, tens into hundreds, hundreds into thousands, etc. But unlike the LoDagaa, who actually bring physical objects together into groups, then form groups of these groups, and so on, the child counting their hoard of marbles does not need to gather ten of them into a pile and ten more into another pile, etc. Rather the groupings are built into the number system. And the child may have learned the number words and learned to count without having grasped the multi-leveled grouping process the system embodies (as is evidenced by the difficulty so many people have with exponents and the concept of orders of magnitude).

The modern reader is likely to view the stepwise procedure employed by the LoDagaa as a general method that might be used to count (in effect) anything. And, of course, it can be used as an abstract counting system, applicable to any collection of things, no matter what their nature, as long as they can be moved about and collected into groups, and groups of groups, etc. But the LoDagaa did not seem so to regard it. Rather, they used it to perform a particular task, to accomplish a specific goal in a concrete cultural setting. For a different task in a different context, e.g., counting cattle, they used a different procedure, as would be deemed appropriate by anyone who has ever attempted to count cattle milling about in a pen or corral.

The important point here is that counting procedures come into being and evolve in concrete circumstances, and their characteristics vary accordingly. It can be seen from the examples cited above that the size of the groups formed can vary. (Basic groupings of four, five, and ten were used.) And groups having a certain number of members may themselves be a member of a group (of groups) having a different number of members, as exemplified by the LoDagaa's collecting cowries into groups of five, then bringing four of these groups together in the next stage of the process. It is no surprise that counting procedures that grow up independently of each other employ different groupings at one or more levels or "ranks."

And, indeed, it turns out (as will be seen below) that at an early stage the Sumerians not only used different procedures for keeping up with different kinds of things, they also made use of a variety of ways of grouping things. But investigators did not reach this conclusion quickly or easily. Rather, being familiar with the sexagesimal number system (with a base of 60 rather than 10) later used in cuneiform writing (long since deciphered), and having found that some of the impressed signs on the clay tablets could be interpreted as numerical signs for sexagesimal numbers, they quite naturally concluded that such was the case. As the result of the general acceptance of this interpretation, the actual, much more complex situation did not come to light until a completely new, computer-aided study of the very early Sumerian clay tablets was carried out late in the twentieth century. Unlike the subsequent proto-cuneiform tablets, which bear incised as well as impressed signs, the **earliest tablets** exhibit only impressions—and just two basic types of these. There are round indentions (holes), presumably produced by holding a round stylus perpendicular to the surface of the wet clay and tapering round hollows (which some writers refer to as "notches" or "wedges"), presumably made by pressing the stylus obliquely into the clay. By using styluses of two different diameters, the Sumerians also made two sizes of each of these two kinds of impressions. According to the sexagesimal interpretation, the smaller tapering shallow hollows represent units, the smaller round indentions or holes represent groups of ten, and the larger tapering shallow hollows six groups of ten or sixty units. And since these three kinds of impressions had been matched, in the way noted above, with small conical tokens, small spherical ones, and large conical ones, respectively, it was inferred that such was the case with them as well.

Early on, however, the sexagesimal interpretation of the impressed signs on the proto-cuneiform tablets ran into problems. As work on deciphering the tablets progressed, investigators recognized that the symbols on the reverse of some of them constituted summations of the information recorded on the front. The problem was that certain sums thus recorded simply did not add up. One suggested solution to this puzzle was that calculative errors on the part of the ancient scribes were responsible for these discrepancies. Apparently, no one considered the possibility that the groupings employed by the Ancient Sumerians might have varied from one context to another and that the symbols for groups (whether small spheres or small round holes, large cones or large tapering hollows) would have represented groups of differing sizes depending upon the context. It is not surprising that this possibility was not considered, as it would seem to imply that at one time the Sumerians had more than one number system, more than one way of counting. As improbable as that may seem, the actual situation may very well be even further removed from contemporary conceptions and presumptions.

As an alternative to projecting subsequent ways of thinking and doing upon the early Sumerians, it is helpful to consider how people accustomed to even earlier and simpler ways might have addressed the need for more effective methods of keeping up with things and preserving information. Presumably, at one time people living in the ancient Near East just used collections of pebbles or tokens in the way described above for such purposes; but as the number of things that they needed to keep track of increased, such collections would have become increasingly unwieldy and burdensome. In such a context, it clearly would have been very advantageous to substitute a token of a different type or larger size for an entire group of the original tokens or pebbles. Furthermore, as things progressed, it also would have been advantageous to substitute yet another type of token for a group of tokens of the second type, and so on. The correspondence between the contents of some of the clay *bullae* and subsequent numerical records leaves no doubt that such multi-level grouping procedures were well established by the time the clay envelopes came into use. However, it is important to note that the use of these grouping techniques presupposes neither verbal counting nor a number system as currently understood. Nor, as seen in the case of the LoDagaa, need the groupings at the different levels be the same. And, most importantly for present purposes, there is no reason to assume that grouping practices that were developed in one area of practical activity would match those devised in another. Thus, while there may be good reasons to believe that, in some cases, small spherical tokens signified groups of ten, it does not follow that they always did.

If the early impressed signs are physical records of information expressed in terms of a variety of grouping systems, then one would expect—given the availability of only a very few distinct types of symbols—that the **same impressed sign** might very well have been **used to represent groups of different sizes, depending upon the context.** And that is exactly what appears to have happened. When a small round impression (hole) was used to symbolize a group of discrete individuals such as people, farm animals, or tools, early investigators were correct to conclude that the group consisted of ten items. But they were mistaken in assuming that such was the case when that symbol was used to designate an aggregate of certain small (volume) measures of grain. The situation with regard to area measures of land and others as well is similar. Thus, the solution to the problematic-sums puzzle lay in recognizing that the early Sumerians used a variety of grouping systems and that an impressed sign might have a signification in the context of one of them that differed from what it would have had in another. Certainly, such a possibility is not something that anyone accustomed to number systems like that learned by young children nowadays could be expected to consider. Thus, it is not surprising that the early investigators erroneously assumed that the impressions on the very early clay tablets were numeric signs, symbols for numbers, each having the same meaning wherever it appeared, rather than being alternatives to three dimensional tokens or *calculi* corresponding to individuals or to groups (or to groups of groups) of one size or another depending upon the context. The impressed signs on one of these tablets did provide quantitative information, but it need have amounted to no more than the existence of a one-to-one correspondence between certain items of interest (or groups of them) and the impressed signs on the tablet.

Since the Sumerians used each of their grouping systems for only certain purposes, they could tell, on the basis of which one of these had been used, together with the overall context, just what kind of commodity a document was concerned with. The early clay tablets with their impressed signs thus had a dual function, providing information regarding the nature of the items involved as well as the quantity thereof. Expressions that convey both quantitative and qualitative information are used by people who have fully developed number systems but play a more significant role in cultures that do not. For example, when used in a context in which it is clear that "biological twins" is intended, the word "twins" conveys both the kind of thing being referred to and the number of individuals involved. And there can be series of words or symbols for groups of a specific kind of thing with memberships of one, two, three, four, five, etc., such as "solo," "duet," "trio," "quartet," "quintet," etc., as applied to musical performers or compositions. Such series are referred to as "concrete number sequences." Another expression used is "numerical classifiers." There is one feature of the series just cited that is not shared by very early concrete number sequences: its terms are derived from pre-existing number words (Latin, in this case), whereas there is reason to believe that the terms of the very early sequences are not.

Despite the difficulties inherent in any attempt to reach conclusions about the linguistic practices of non-literate peoples (especially in the remote past), there is evidence of the widespread utilization of concrete number sequences when conditions make their use advantageous. The Fiji Islanders, for example, call ten boats *bola* but speak of ten coconuts as *koro*. In an anthropological study conducted in the late 1880s, Franz Boas found that the Tsimshian, one of the Northwest Coast peoples whose complex way of life would have occasioned the development of methods of numeration, had a separate series of number words for each of the following classes: men, canoes, long objects (such as trees), flat objects (such as blankets or robes), round objects, measured quantities of substances, and finally things not included in the first six groups. For example, they referred to five men as *kcenecal* and five canoes as *kctoonsk*. Ten men was rendered *kpal* and ten canoes as *gy'apsk*. The Gilyak of the coastal region of Asia just to the north of Japan employ twenty-four different sequences of concrete number words. The forms that these concrete number sequences take are extremely varied. In some series one finds variations of root expressions but not in others. In some cases the same affix appears in more than one sequence, while in some instances there is no obvious linguistic connection between the members of one series and those of another used by the same people. Such variability is, of course, just what one would expect in the absence of number systems. And although concrete number sequences are not just way stations on the road to elementary mathematics as we know it, they are an important part of the environment in which number systems were developed.

As the Sumerians' economic and administrative activities became more and more complex, they expanded their repertoire of record-keeping techniques. Along with the usual impressed signs, **they began to inscribe symbols** in the wet clay with a **sharp stylus**. The earliest examples of these show up in the archaeological record about two hundred years after the first clay tablets with impressed signs. These inscribed symbols appear to represent agricultural products and other commodities. Presumably they were introduced to address the need to record more specific information concerning a matter than was possible with the older system alone. Some of these signs are **pictographic**; others are quite abstract. For example, an inscribed pinnate image has been determined to be a symbol for barley; a circle enclosing a cross (two diameters intersecting at 90°) stood for sheep. A striking feature of these inscribed symbols is their well-defined nature and the uniformity with which they appear on tablets. This characteristic is, of course, fundamental to any system of writing. Consider the effort required to develop handwriting that is "legible." The appearance of the Sumerians' inscribed symbols marks the earliest stage of one of the most profound developments in human history: the invention of writing.

B. Proto-Cuneiform and the Sounds of Human Speech

Some of the earliest **tablets** bearing both incised and impressed signs also had been marked in the usual manner with a cylindrical seal. **Three kinds of information** are recorded on these tablets: (i) the **type of thing** being dealt with (e.g., sheep, measures of grain, units of labor, etc.), (ii) the **number** or quantity thereof, and (iii) the **identity of** the **person** or persons involved. However, a record-keeping system of this kind has serious limitations. To perform their function, cylindrical seals must be both unique and generally recognizable, two characteristics difficult to achieve simultaneously in a complex, large-scale society. Accordingly, it can hardly be surprising that very few clay tablets have been found that are marked in the way just described. Once the Sumerians had begun to inscribe signs on the wet clay tablets, little time elapsed before impressed seals were supplanted by inscribed signs as a way of identifying persons. This step marks the advent of "proto-cuneiform," a writing system for keeping records of the three kinds of information listed above. It provided the foundation for the cuneiform writing of the Sumerians, Akkadians, Assyrians, and Babylonians. Since Sumerian names did not lend themselves to pictorial representation, the bulk of these very early signs could hardly have failed to be arbitrary and conventional. Since there is no obvious way to determine whether a conventional sign represents a person or a product, modern linguists have had great difficulty interpreting these early clay tablets.

The overall approach taken by the Sumerians to the problem of recording the names of individuals was to **devise a sign for each of the sounds that occurred** in a Sumerian name. By means of these signs, names could be recorded phonetically. And while present knowledge is incomplete and short on specifics, the basic nature of the procedure they followed is quite straightforward. It consists in **matching names or syllables** which occur in names **with like-sounding words** that do lend themselves to **pictorial representation.** Thus, to give a contemporary example, one might represent the name "Gayle" by a sketch suggesting things being blown about in a high wind or "gale." A strategy of this kind was highly suited to the Sumerian language because it is predominantly monosyllabic, has numerous homonyms (words that sound alike), and is agglutinative in nature (retains the forms and meanings of linguistic units, "morphemes," when they are joined into compounds). Under these conditions, it was not too difficult to match a given syllable with a like-sounding word that could easily be represented pictorially. By taking this step, the Sumerians laid the foundations for the world's first system for recording human speech, but this process took time, as did the emergence of a number system as is currently understood.

Viewed retrospectively, the markings on the very early proto-cuneiform tablets seem to consist simply of impressed numerals and inscribed signs of two sorts: those identifying individuals and those identifying commodities or administrative matters. But if one focuses upon the challenges facing the Sumerian bookkeepers and the strategies they adopted in addressing them, the picture becomes more complex. It is important to keep in mind that the purpose of proto-cuneiform documents was to preserve information absolutely vital to the Sumerians' economic and administrative activities. Thus, as a practical matter, the documents contained no more information than was needed for the matter at hand, omitting anything that was peripheral or could have been assumed to be known. Furthermore, five quite distinct numerical sign systems (and several variations thereof), reflecting the different grouping systems in use even earlier, can be found on the early proto-cuneiform tablets. Since, as explained above, each of these systems was used only for certain kinds of things, the very fact that a particular one was used conveys some information about the subject of a given proto-cuneiform record. Similarly, information regarding the nature of a product or commodity, as well as the quantity thereof, would have been relevant to establishing the identity of the human agent(s) involved. Thus, the **symbols on the early proto-cuneiform tablets functioned in concert rather than separately and independently.** As used, the early proto-cuneiform symbol system did not maintain a clear distinction between the quantitative and the qualitative aspects of things.

C. Early Number Systems and Cuneiform Symbols

Several centuries would elapse before the Sumerians' specialized numerical systems would be replaced by a single system free of involvement with qualitative considerations and equally usable in all kinds of contexts. The steps taken in this process bring into clear focus the essential elements of number systems like that currently taught in elementary schools. Since one of the numerical sign systems found on the proto-cuneiform tablets evolved into the sexagesimal system used by the Akkadians, Assyrians, and Babylonians, it is a good place to start. Early on, this system took the form of a multi-leveled grouping procedure like that used by the LoDagaa, except it used alternating groups of 10 and 6 rather than

5 and 4. And, of course, it used impressed signs to represent groups rather than actually forming them physically. In addition to symbols for units, for groups of 10, and for 6 of those groups (60 units), as described above, the proto-cuneiform tablets (in time) bore symbols for 10 groups of 60 (a large tapering shallow hollow with a small round indention impressed in it), for 6 groups of these groups of 600 (a large round but shallow indentation), and for 10 groups of 3,600 each (a large round but shallow indentation with a small round indentation in its center) representing 36,000 units.

In order to illustrate the variety of grouping procedures employed by the scribes who produced the proto-cuneiform documents and to gain insights into the way number systems function and develop, a second numerical sign system will now be considered. In contrast to the numerical sign system just described, this second system includes only six individuals or units in its most basic group; but it combines ten of these groups in its second-rank groups (encompassing sixty units) and, in turn, three of these into its third-rank groups (each encompassing 180 units). Apparently, the basic units of this system corresponded to some commonly used measure of grain, and the first-rank groups to a quantity of grain six times that large, and so on. The system thus parallels the English system of dry measure, in which two pints make a quart, eight quarts a peck, and four pecks a bushel. In general, it appears that the different Sumerian numerical sign systems derive from the separate and independent procedures used by the early Sumerians when engaged in such tasks as recording the size of groups of sheep, keeping track of quantities of grain, specifying the size of plots of land, and so on.

As Sumerian society became more complex with the passage of time and the need to bring together different kinds of quantitative information, the specialized numerical sign systems were gradually supplanted by the first system just described. But this did not happen all at once. Rather, the changes involved took place in fits and starts and at different times in different locations. Few details of this process are known, but by about 2500 BCE only two basic numerical sign systems remained in use: the numerical sign system from which the sexagesimal number system evolved and a specialized system used only in connection with areas of land. During roughly this same period, i.e., beginning around 3000 BCE and continuing through most of the third millennium BCE, a new method of inscribing signs on clay surfaces came into use. Instead of either a cylindrical or sharp-pointed stylus, scribes began to employ an instrument with a triangular cross-section. With it they made two basic kinds of marks on clay surfaces. One of these is usually described as "wedge-shaped" or "cuneiform," but it more closely resembles the outline of a golf tee. The other, less well known and less used, has been called a "hook"; it resembles a boomerang or (sweptback) flying wing. As the reader probably suspects, the old wedge-like tapering hollows were replaced by the new "**cuneiform**" marks (see Figure 6-2 below). Initially, there were two sizes of these: the smaller one served as the numeral "1" and the larger one (signifying 60) was somewhat asymmetrical. Small round holes (which stood for 10) gave place to "hooks," the impressions resembling boomerangs. Hooks were combined with the larger wedges to form symbols for 600, previously signified by a large tapering hollow with a small round hole in it. And, since curved lines could not be accommodated in the new system, the large round impressions standing for 3,600 were rendered as roughly diamond-shaped figures formed by four wedges. Rather unremarkable in themselves, these innovations set the stage for some truly exceptional developments.

Figure 6-2. Simple to Complex Cuneiform. The bottom row shows very simple wedge marks on tablets (left) to increasingly complex cuneiform (moving right). The top row shows more clay envelopes (*bullae*) on the left and tokens to the right.

Following the drawn-out transition from curvilinear pictographs and rounded impressed numerals to the cuneiform system, there was a burst of mathematical innovation in the comparatively short period between 2100 and 1800 BCE. Little or nothing is known concerning how these advances came about; however, on the basis of documents known as "Old Babylonian School Texts" from roughly 1800 BCE, it is clear that by then a sexagesimal **place-value number system** had been worked out. (By this time, of course, lower Mesopotamia was being brought under Babylonian rule.) In the older Sumerian systems, each symbol (except those denoting units or individuals)—whether it be a tiny sphere, a small round hole in a clay tablet, or a tiny boomerang-shaped imprint—stood for a specific grouping, having a definite rank and number of members. (Each of the symbols just mentioned signified a group of ten units or individuals, a rank-one grouping.) Multiples of a given group (and multiples of units) were indicated simply through repetition. Thus, the number thirty-three was expressed by making three hooks (three boomerang-shaped impressions) and three tiny wedges. As time went by, it became the practice to inscribe the symbols expressing a number in descending rank order (from left to right). While not strictly necessary, this practice was surely an aid to speed and accuracy, and it paved the way for the place-value approach.

In a **place-value system**, groupings and ranks are not identified by special symbols but rather by places in an order of rankings. The Babylonian system had a place for units and next to it, proceeding from right to left, a place for rank-one groupings, then a place for rank-two groupings, and so on. It thus had the same structure as the Indian ("Arabic") number system, learned by every schoolchild, in which the ten's place is just to the left of the unit's place, the hundred's place just to the left of that, and so forth. Thus a "5" may signify 5 units, 50 units, or 500, depending on its placement. In contrast, the Romans used "V" to signify 5 units, "L" to signify 50, and "D" to signify 500. (Roman numerals are not a place-value system.) The Babylonian system was also like the Indian system in that both use the same grouping method at every level: there are the same number of rank-one groups in a rank-two group as there are units in a rank-one group, and so on. Only when the same grouping method is used at every level is it possible to speak meaningfully of a "base." In the Indian decimal system (so called from *decem*, Latin for ten), every group (of every rank) is formed of ten members of the next lowest rank: a hundred consists of ten groups of ten units each, a thousand consists of ten groups of a hundred each, etc. Instead of ten, the Babylonian system is constructed around the number sixty. Its rank-one groups encompass sixty units, its rank-two groups encompass sixty of the rank-one groups, and so on. Using the same grouping method at every level is a significant departure from the old Sumerian system, which, as explained above, used alternating groupings of ten and six. Thus, the Sumerian system was not "sexagesimal" in the strict sense. Alternating (and varying) grouping methods like those discussed above probably have advantages in non-literate societies; but, as will be seen below, the Babylonian place-value system, using the same grouping procedure at all levels, brought exciting new possibilities into view. Of course, the new system had to have numerals from one through fifty-nine (rather than just through nine!) which could be inscribed in the unit's place, in the 60's place, in the 3,600's place, and so on. For this purpose, the Babylonians used combinations of hooks and wedges of the type described above. Accordingly, the numeral "thirty-three," three hooks and three wedges, could be used to record thirty-three units or thirty-three groups of sixty, and so forth depending upon where it was placed. Grouping things into tens (represented by a hook) thus survives in a secondary role in the Babylonians' sexagesimal system.

One of the most significant characteristics of the new place-value system is its open-endedness, a feature of number systems now taken for granted. In such systems, there is no "largest number." The introduction of the place-value approach thus constitutes a fundamental departure from the old Sumerian system, in which a finite set of symbols each stands for a specific quantity or number; one of these must be larger than any of the others and thus the largest number of the system. This does not mean that, prior to 2100 BCE, the Sumerians were unaware of the possibility of things being so numerous as to exceed the bounds of their number system; but it does mean that beyond that limit they were without a means of ordering things as regards size or number, that they were in a situation not unlike that of the South African Khoisan when confronted with "many" individuals or things. But unlike the Khoisan, the Sumerians had much practical experience with grouping and by the end of the period had extended the grouping process by two more steps: an additional grouping of ten and another of six (following the alternating grouping pattern noted earlier). The second of these steps was taken only as the transition to cuneiform was underway. One may reasonably assume that the possibility of applying the grouping process again and again was on the minds of the Sumerians who introduced the place-value number system.

A place-value system offers tremendous computational advantages. Evidently, the early Babylonians were aware of some of the possibilities offered by the new system and had set about exploring them. Inscriptions on some of the tablets recovered are sexagesimal versions of multiplication tables, perhaps produced as practice exercises or maybe for reference. There is also clear evidence of extensive work with reciprocals and fractions and even solutions to such problems as finding the dimensions of a (rectangular) field with a given area and perimeter problems lacking practical application and of theoretical interest only. Indeed, accounting of all kinds in Old Babylonia continued to be done using the old Sumerian system. Thus, the advantages of the new, place-value system were not really exploited until some 1,200 years later in a latter-day, revived Babylonian or "Chaldean" empire, which reached its zenith under Nebuchadnezzar II (who ruled from 605 to 562 BCE). The exploitation of Old Babylonian mathematics by the Chaldean astronomers is one of the earliest landmarks on the road to modern science.

The Old Babylonian system did have some serious shortcomings, however, perhaps most notably the lack of a zero. Thus, its users would have had to leave blanks where appropriate and remember their significance later—not an easy assignment, as is demonstrated by an error in a surviving Old Babylonian table which resulted from the failure to follow through on just this point. Evidently, the Chaldean astronomers were aware of this problem and devised a way of addressing it—a small place marker to indicate that no groupings of a particular rank were included in a given number. For example, the appearance of this small symbol between a wedge and a hook would signify that no groups of sixty were included in the number, just one group of sixty squared plus ten units ($1 \times 60^2 + 10 = 3{,}610$) rather than one group of sixty plus ten units ($1 \times 60 + 10 = 70$). However, the Chaldeans used the place marker only between two other symbols and not to the right of them, as is done with the zero in 20 or 200. Thus users of the system were left to determine from the context whether a wedge was to be read as 1, 60, 3,600, and so forth. A further step toward a true zero may be found in the writings and calculations of the Greek astronomer Ptolemy (second century CE), who used a small symbol, "*o*," in writing angular quantities as Babylonian sexagesimal fractions (the familiar system of degrees, minutes, and seconds!). He used this symbol to indicate both the absence of a fractional group, as in 29° o' 25", and the absence of a whole unit, as in o° 25' 23". But the familiar zero, an essential part of the powerful place-value system now so widely used, did not appear—as one of the Indian numerals—until six centuries later.

D. From Record Keeping to Writing

During the same period that the ancient Sumerians were simplifying and unifying their cumbersome number systems, they were taking great strides in other aspects of record keeping. Most significantly, as described above, they initiated the practice of using **symbols to represent** the **sounds of human speech**, specifically for the purpose of recording the names of individuals. Contrary to what one might expect, the introduction of phoneticization was not followed by the development of a complete syllabary for the Sumerian language (i.e., a symbol for every distinct syllable in the language) or anything close to it. Nor was it followed by the wholesale abandonment of the abstract logograms (symbols for words) and pictograms already in use. Simply put, **Sumerian cuneiform** is not a systematic way of recording human speech. It is rather a complex conglomeration of devices introduced to address the mounting record-keeping challenges faced by the Sumerian scribes—a system which eventually came to have the capacity to represent human speech. Sumerian cuneiform embraces **three** quite distinct **kinds of signs:** (i) **logograms** (cuneiform renditions of both pictograms and abstract symbols), (ii) **syllabograms** (symbols for syllables), and (iii) **determinatives** (non-phonetic signs further specifying the intent of the scribe). As a result of the way Sumerian cuneiform was developed and used, it has characteristics that can hardly fail to confuse and frustrate modern scholars attempting to decipher it. For example, the symbols representing syllables are also used to represent words—the words for which they were symbols to start with. Faced with one of these symbols, the modern scholar thus has the added burden of deciding whether to treat it as a word or as a syllable. But this dual role posed little problem for the Sumerian scribes, who could easily tell from the context which use was intended. Reliance like this upon the reader's ability to supplement the information provided by a document was characteristic of Sumerian record keeping throughout the period. For practical reasons (the time required to write, the bulkiness and weight of the clay tablets, etc.), the earliest records included only an absolute minimum of information (as seen above). In this context, it can hardly be surprising that, in the earliest cuneiform documents, one finds recorded only the nuclei or "stems" of words—without prefixes, suffixes, or other additions which distinguish nouns from verbs, identify the tense and person of verbs, fix the number and case of nouns, etc.

At some point, of course, the Sumerian scribes did begin to use syllabic signs to express such grammatical elements. Since it is so difficult to communicate such things as tense, number, and case pictorially, and since the capacity to express them is so vital to a fully written language, one might have expected that it was precisely this limitation of pictographic expression that actually prompted the development of phonetic representation. But such seems not to have been the case. First of all, it is clear from the archaeological record in southern Mesopotamia that recording personal names by means of syllabic symbols preceded the use of such symbols to express grammatical elements. Furthermore, when the Sumerians did begin to utilize syllabic symbols for such purposes, they proceeded in a piecemeal fashion, expressing one grammatical element phonetically at one time and additional ones only after various periods of time had elapsed. Indeed, affixes (prefixes, suffixes, and infixes) were not specified fully until early in the second millennium, when Sumerian probably had ceased to be a living language and survived only in the schools, rather like Latin during the Middle Ages. That phoneticization was introduced to facilitate the expression of grammatical factors thus seems highly unlikely.

The omission of grammatical detail and the dual role of syllabograms are not the only source of ambiguity in Sumerian cuneiform. The limitations of Sumerian syllabaries also give rise to problems of this kind. These systems simply do not provide a way of distinguishing certain types of consonantal sounds from others (i.e., no way of discriminating between voiced, voiceless, and emphatic consonants occurring at the end of syllables). And some signs stand for any one of a series of syllables, the same consonant followed by any vowel. In some instances in which needed syllables were not included in their syllabary, Sumerian scribes simply substituted signs for syllables with similar consonants. Given the importance of record-keeping systems in complex societies, one would certainly expect the Sumerians to have developed ways of clearing up ambiguity when the need arose. And, indeed, they had an entire repertoire of notations used by scribes to make the intent of their inscriptions clearer. These "**determinatives**," as scholars refer to them, have no phonetic role (no direct connection to language as spoken). They rather clarify the meaning of inscriptions by identifying the class to which items referred to belong or by indicating which among the possible phonetic values of symbols are the intended ones. These two functions of determinatives, one semantic and the other phonetic, reflect the mixed character of Sumerian script. The workings of these devices can be illustrated by means of the cuneiform symbol for the word stem (or nucleus) for "plow" or "cultivate." If preceded by the determinative signifying "wood," it would have been read "plow," but if preceded by the semantic indicator for "man" (or simply with no determinative), it would have been read "plowman" or "farmer." Other such semantic indicators signal that a symbol stands for the name of a deity, that of a woman, a city or place, etc.

Another technique used by the Sumerian scribes to overcome the limitations of their syllabaries is "**syllable-telescoping**," as it is termed by linguists. Basically, it consists in using the (combined) symbols for two short syllables to represent a longer one, with the sound of one of the short syllables beginning as does the sound of the longer syllable and the sound of the other short syllable ending like the longer one ends. In addition, the two short syllables are chosen so that they otherwise sound like each other and like the middle of the longer syllable. (Other variations are possible.) Using, in this way, the symbols for two shorter syllables to stand for a longer one results in its middle sound being represented twice. It is from this "overlap" that the name "telescoping" derives.

In order to describe more precisely the specific technique utilized by the Sumerians, it helps to employ the concepts of "vowel" and "consonant." However, the reader should not assume that these concepts—or the distinction they mark—were in play in ancient Sumer. More time would elapse before the differing aspects of human speech which they represent would be recognized or understood. As noted earlier, Sumerian words are predominately monosyllabic. And, aside from singular vowels, they assume three basic forms: a vowel preceded by a consonant (CV), a vowel followed by a consonant (VC), and a vowel both preceded and followed by a consonant (CVC). Although syllables of the third type are quite numerous, Sumerian syllabaries contain syllabograms for only a few of them. For the remainder, Sumerian scribes used syllable-telescoping. To represent a syllable of the form C_1VC_2, they combined the symbols for the syllables C_1V and VC_2. For example, "ral," for which there is no separate syllabogram, would be written as "ra-al."

For anything even approximating a written language to emerge, conventions of various sorts must be established. Most fundamental of all, the forms of the symbols used must be sufficiently standardized to provide for reliable recognition. Achieving this end becomes more difficult and more important as symbols become more abstract and more stylized. Success requires concerted effort over an extended period of time. Ancient documents from early in the second millennium BCE testify that mastering the cuneiform script was a long, wearisome process. Extensive schooling for at least a fraction of the population had arrived! And there is archaeological evidence of student labors as early as 3000 BCE. This includes nondescript lumps of clay across which lines have been incised and/or rows of simple impressions made with a stylus. A second group of clay objects believed to be the remains of ancient practice exercises encompasses everything from amorphous lumps to tablets of the standard form used by scribes throughout Babylonia. The members of this group differ from those of the first group in that they are inscribed with recognizable Sumerian symbols or parts thereof. But the inscriptions on them have neither the arrangement nor all the elements of an administrative text. Repeated versions of the same symbol on some of them bear mute testimony to long-ago efforts to produce acceptable script. Lumps of clay with partially obliterated inscriptions appear to be ancient analogues of wadded up pieces of paper.

Most significant is a third group of clay objects usually referred to as "**school texts**" or "**lexical lists**." Aside from administrative/economic records, which make up the overwhelming majority of early Mesopotamian documents, lexical lists are the most numerous, comprising some 15 percent of the total. The organizing principle of these word inventories, while not entirely clear in every case, is to gather words for the same kinds of things into lists. There are lists of animals of various kinds, of trees and shrubs, of wooden objects, of pottery vessels (with varying contents), of place names, etc. The categories employed correspond closely to those utilized with the determinatives. Indeed, each entry in the list of wooden objects includes a determinative, the symbol for "wood." Some of these categories are divided into subclasses. Animals, for example, are subdivided into cattle, fish, birds, etc. In some cases, males are distinguished from females and juveniles from adults. Thus, organizing things on the basis of types and subtypes (rather than around the tasks and concerns of daily living) is characteristic of literate cultures, as was noted above (in connection with the development of numbers and counting) and as will be discussed more fully following this case study. The example of the Sumerian school texts is especially instructive in that it suggests a very concrete explanation for this particular feature of literate societies. Whether these tablets are viewed as practice exercises or as reference works rather like dictionaries (as some scholars urge), it is clear that such organization and structure hold significant advantages. Thus, their presence is the result, at least in part, of strategies adopted by those struggling with the challenges of learning and using writing systems. That much emphasis was placed upon precise replication is attested by the presence in the archaeological record of school texts meticulously copied over a span of eight centuries. The existence of these multiple copies is, of course, one of the things that distinguishes school tablets from administrative documents.

As remarkable an achievement as Sumerian cuneiform was in itself, its overall significance stems from its having been taken over for use with other languages and from its pioneering role in the process of phoneticization, the representation of the sounds of human speech. In the first such appropriation, the script was used to write Akkadian, a Semitic language spoken by the Babylonians, who took control of lower Mesopotamia during the latter part of the third millennium BCE. The Babylonians adopted many Sumerian cultural elements, including the Sumerian writing system—even though it is not well suited for writing a (non-agglutinative) Semitic language like Akkadian. Rather than adding suffixes or prefixes to unaltered cores or roots (as is done in the Sumerian language), Semitic languages distinguish nouns from verbs, subjects from objects, verb tenses from one another, etc. by means of vowel changes. Examples of this same linguistic pattern can be seen in modern English, viz., verbs with principal parts like "sing," "sang," and "sung" and nouns with plurals like "geese" (singular, "goose"). This feature of the Akkadian language made it essentially impossible for the Babylonians to adopt the same basic strategy that the Sumerians had followed: first specify the relevant stem or root, then add as much (or little!) grammatical, phonetic, or semantic detail as is necessary to achieve the desired result. (See the above example of "plow" or "cultivate.") To specify a particular Akkadian stem or root without at the same time spelling out additional grammatical detail (such as the tense of a verb), one would have to have a way of writing consonants separately from vowels, such as is found in the Greek and Roman alphabets as well as in such consonantal scripts as Hebrew and Arabic. (In the case of the above example, such a capability would mean being able to write "sng," with no vowel, leaving the tense of the verb unspecified.) The inhabitants of Ancient Mesopotamia, however, had no such way of writing. Nor is there any reason to think that they made any systematic effort to analyze the spoken word into units smaller than syllables. What the Babylonians actually did was to match the syllables of their own language with Sumerian ones and to use the Sumerian syllabograms, as well as some of their own logograms, to write the Akkadian language.

Perhaps most significantly, the Babylonians and others who availed themselves of the cuneiform script carried forward the process of phoneticization and at the same time evolved simpler and more systematic ways of writing. As seen above, Sumerian cuneiform is a complex system, fundamentally logographic but including a set of determinatives or clarifiers and a syllabary. The syllabary made it possible for words to be recorded phonetically, i.e., according to how they sound. In contrast, Old Babylonian is primarily a syllabic system which also employs a few word signs (logograms) and clarifiers. Similarly, the cuneiform writings of the Hittites of Asia Minor are primarily syllabic in character, almost exclusively so in the later periods.

E. The Literacy Revolution

The great advantage of syllabic systems is that they require a much smaller number of different symbols than does a logographic (or ideographic) system. Systems with a significantly smaller number of symbols are, of course, easier to master and are thus more likely to be learned and used by a larger fraction of the population (other factors being equal). An even more significant feature of syllabaries (as compared to logographic systems) is that, once set up, they provide a means of recording elements of speech not previously written down, without introducing additional symbols. Because of the effort required to get enough people to learn a new symbol for it to be useful and because of the burden that a system with numerous symbols imposes upon its users, new symbols are likely to be introduced only if doing so provides a significant advantage. On the other hand, if a system of symbols offers, as a kind of free bonus, a means of representing linguistic elements not previously provided for, then not only are users of the system likely to develop symbolizations for more of the same kinds of things they already have symbols for (they would be able to write anyone's name—not just those of a relatively small number of people, as in the case of impressed seals), but they are also likely to develop ways of symbolizing things (e.g., verb tenses) that previously were not symbolized ("written") at all.

As people's capabilities for making records and communicating by means of physical symbols developed, they found more and more applications for this powerful technology; these include not only records of administrative and commercial transactions, but also governmental proclamations, legal codes, and inscriptions commemorating the exploits of powerful rulers as well as historic events. Soon there were letters and written versions of oral narratives, ballads, etc. There are at least half a million cuneiform tablets from throughout Mesopotamia estimated to be held by the world's museums, and many more were produced, and while the majority are administrative or economic documents, a significant proportion were used for other more creative or intellectual purposes. For example, at least 10 percent of the tablets recovered from the Sumerian city of Uruk are non-administrative. Eventually, as fully literate cultures emerged, so did sacred writings, epic poetry, lyric poetry, drama, essays, history, philosophy, and science along with mathematics. While it is easy to understand that writing is necessary for these human endeavors, it is not easy to grasp the significance of this innovation. As explained in the Introduction, the problem is that **literacy changes the way people think**; and unless one has a basic understanding of non-literate cultures, one cannot comprehend the difference that literacy makes.

The first steps toward a better understanding of oral cultures were taken by Milman Parry and Albert Lord who made pioneering studies of Homeric and Balkan narratives. Of special relevance for present purposes is the field work carried out by A.R. Luriia among non-literate peasants in remote areas of Uzbekistan and Kyrgyzstan in the very early 1930s. The results of these studies were not published until 1974 (in Russian). In starkest terms, Luriia found that when identifying, classifying, and relating things, these peasants did so in terms of common objects and practical situations and tasks rather than in terms of geometrical shapes and abstract categories like "tools" or "plants." For example, a drawing of a circle would be identified as a plate, a square as a door or drying board. When asked which object does not belong in a group consisting of a hammer, a saw, a log, and a hatchet, they did not, as literate people do, select the log (because only it is not a tool) but (as revealed in conversations with Luriia) focused, in making their decisions, upon various aspects of carpentry in which wood plays a central role. A related phenomenon which Luriia encountered is that non-literate people do not, in general, make a practice of defining words by means of other words. Thus, not only are there, of course, no dictionaries, but there is, by and large, no oral analogue of what dictionaries are used for. When pressed to explain what even the most common objects are, Luriia's subjects clearly found the request strange and were reluctant even to attempt an explanation. Rather, they questioned why anyone would want to do such a thing when everyone knows what these objects are. Faced with the query, "If you went to a place where there are no cars, how would you explain to people there what a car is?" they gave responses that evidence the limitations of their conceptual resources at the needed level of abstraction, the absence of such terms/concepts as "conveyance" and "vehicle."

A word of caution is in order. Nothing in the preceding account or in what follows should be interpreted as suggesting that non-literate people are not intelligent, not wise, or incapable of giving explanations. The unreflective application of literate standards to oral cultures must be guarded against, not only out of respect for the non-literate but in order to grasp the significance of the invention of writing.

Studies by Luriia and others have brought to light the challenges faced by those who first attempted to develop systematic (as opposed to narrative) accounts of the world, its structure, and dynamics: the lack of an appropriate vocabulary (set of categories) in which to formulate such accounts. Eric Havelock, another of the pioneers in the field, addresses this topic in his "The Linguistic Task of the Presocratics, Ionian Science in Search of an Abstract Vocabulary" (1983), which the interested reader may wish to consult. As noted in Chapter II, crucial to the work of Isaac Newton was the (prior) development of the concept of motion that is accelerating at a constant rate. Non-literate patterns stand in sharp contrast to Aristotle's stress on the importance of definition in terms of genus (general class) and differentia (specific difference) and his systematic, hierarchical treatment of concepts in the *Categories*, a way of thinking that literacy made possible.

Even more intriguing to anyone who has ever studied geometry or formal logic are the reactions of these non-literate peasants to situations in which a conclusion is clearly warranted, or would be judged so even by persons who have never studied logic (but are literate). Luriia posed the following query to his subjects: "In the Far North, where there is snow, all bears are white. Novaya Zembla is in the Far North and there is always snow there. What color are the bears?" A typical response was: "I don't know. I've seen a black bear. I've never seen any others.... Each locality has its own animals" (Luriia, 1976, pp. 108–09; quoted in Ong, pp. 52–53). Evidently, the person giving this response was not in the habit of according to deductively valid arguments the kind of finality that so impressed Pythagoras, Thomas Hobbes, and essentially everyone else who has ever studied geometry. The difference between the reactions of the latter group to such arguments and those of the peasants is often cited as one of the identifying characteristics marking the orality/literacy boundary.

But scholarly opinion is by no means unanimous regarding the nature and effects of literacy. Indeed, differing views on these questions are to be expected, given the variety of writing systems that have been developed and the differing cultures into which they have been introduced. The practice of writing in China dates from the early second millennium BCE, and an impressive tradition of mathematics was clearly underway by at least several centuries BCE. From this period came, for example, *The Nine Chapters on the Mathematical Art*, which included inductive mathematical proofs, methods for solving linear equations, and applications of geometric principles such as the Pythagorean Theorem. However, texts were generally not concerned with formal deductive proofs (as in Euclid's *Elements*), nor did China give birth to an on-going tradition of reflection and speculation concerning natural entities and processes which paved the way in the West for the rise of modern science. Furthermore, despite its earlier technological pre-eminence (see Chapter VIII), China had fallen behind in that area as well by the middle of the nineteenth century and was unable to resist the encroachments of the Europeans and (later) the Japanese.

A variety of reasons for this complex chain of events have been proposed, many of them controversial. Among the latter is the idea that Western, phonology-linked systems of writing facilitate both the kind of abstract thinking and the type of formal reasoning that Luriia found lacking among non-literate peasants. As noted in the Introduction, these two features of literate thought were essential to the blossoming of geometry and the first stirrings of science in Ancient Greece. Only two centuries or so elapsed between the Greeks' acquisition of an alphabet (by no later than 800 BCE) and their first construction of deductive proofs of geometrical theorems. In contrast, the signs of the Chinese script are symbols for words or, more accurately, concepts (kinds of things). There is no connection between them and the sounds of human speech. Learning to speak one of the Chinese dialects is a completely different undertaking from learning the Chinese script. The Japanese, who speak a language that is quite distinct from Chinese, use the Chinese characters (along with syllabaries) to write their own language. As pointed out above, the big disadvantage of logographic scripts is the number of symbols that must be committed to memory: in the case of Chinese, from a few thousand to fifty thousand depending upon the level of functioning needed. Such a system does not facilitate the spread of literacy among the general population and would not appear to be conducive to conceptual innovation, especially among those whose primary means of communicating is by writing. But whatever the advantages and disadvantages of different kinds of writing systems and however relevant (or irrelevant) scientific work is to technological innovation, many other factors exert powerful influences upon social, political, technological, etc. trends, especially long-term ones. This point is amply illustrated in Chapter VIII by China's abandonment (in the fifteenth century) of its highly successful deep-water naval program and Japan's reversal (in the early seventeenth century) of its initial adoption of firearm technology.

At the core of the literacy revolution is the power of writing to capture, to make available for future use or reference, aspects of things that, in themselves, are ephemeral, some to an extreme degree. A spoken word is no sooner uttered than it is gone. Writing endows linguistic expression with a permanence it would not otherwise have. It achieves this by establishing a link between sound and sight (even if it is an indirect one, as is the case of Chinese script). Prior to the invention of **writing**, people had to rely upon their memories to preserve information concerning past events, proclamations, pledges, wise sayings, the teachings of religious figures, etc. Sometimes well entrenched traditions of relying on memories to preserve accounts of past events, narratives, teachings, etc. have persisted even in societies that have become large and complex. The ruling class in India during the centuries prior to about 500 BCE maintained a strong tradition of the oral transmission of such aspects of their culture. Because of the strength of this tradition and the power of the ruling class that supported it, writing did not make much of a cultural impact when it arrived there during this same period. But changes were on the way, both to the existing order and to the relative importance of writing. The founders of Buddhism and Jainism challenged customary social arrangements and ways of living. Traditional accounts indicate that after the death of the Buddha, the "Enlightened One," his followers employed memorization to preserve his sayings and teachings until they were written down. Indeed, the first evidence of the use of writing in India for other than commercial purposes are Buddhist sources from the fifth century BCE. (Not taken into account here are the undeciphered inscriptions of the Indus valley civilization dating from about 2500 to 2000 BCE. The identity of these people and their language is unknown.)

Writing also greatly facilitates the learning of "foreign" languages. It literally makes it possible to study a foreign language and to translate material from one language to another. While the ability to communicate in more than one language is not unknown among non-literate people, it is not sufficiently common to be a major factor in the spread of the teachings of a religious leader beyond his or her language community. Freed of its dependence upon oral transmission (and a community structure to support it) and accessible through translation, Buddhism spread through southern and eastern Asia. Sacred writings also played a key role in the spread of two other great world religions, Christianity and Islam.

To achieve a more well-grounded understanding of oral cultures and thus of the impact of literacy, it is helpful to consider the challenges faced by non-literate people and their responses to these challenges as they sought to preserve information vital to surviving external threats and disasters as well as those things that are vital to maintaining internal social cohesion. If the information available to a community were confined to what its living members have gained through their own experiences, its store of knowledge would be severely limited, especially with respect to dangers and problems unlikely to be encountered during the lifetime of a single individual or too infrequently for correlations to be noticed. These include such things as flash floods in desert and semi-desert areas, tsunamis, wind-driven fires started by "heat lightening," etc. Thus, the survival of such societies (to say nothing of their prosperity) depends upon their preserving valuable information gathered by their forebears—and it must be done by memory power alone (at least until ways are devised to assist in such endeavors, as discussed above).

While the followers of the Buddha may have memorized his sayings and teachings, the usual practice of non-literate people is to tell and retell stories, not by memorizing them word-for-word but by remembering the crucial parts of the narrative and utilizing, in each re-telling, a basic stock of **mnemonic devices.** It is a characteristic feature of human beings that they relish—and remember—stories. Thus, those ideas, facts, etc. that can be cast in narrative form are more likely to be remembered. Even today, people making presentations involving sophisticated analyses and quantitative data are advised to include a story or anecdote, on the grounds that the latter are much more likely to be remembered and thus will serve as a reminder of other parts of the presentation. In oral societies, narratives are delivered in a **rhythmic cadence** making use of **alliteration, assonance, and repetition.** The structures of the stories tend to fit standardized patterns, and descriptive phrases are formulaic (in ways that would be considered hackneyed by contemporary, literate standards). That such modes of expression facilitate recall is evidenced by the relative ease with which one can learn another person's name if it exhibits one or more of these characteristics, e.g., if the name is familiar or embodies familiar and/or rhythmic ways of speaking. Since remembering certain things is vital to the survival and prosperity of non-literate societies, ways of thinking and speaking that promote this result are reinforced, generation after generation.

Another important characteristic of non-literate thought and expression is the dominance of additivity as contrasted with subordination. Oral narratives typically consist of series of relatively short, independent statements, with the liberal use of "ands" joining them. Much less in evidence is the subordination of one statement or clause to another. Such asymmetrical relations are found in complex sentences with introductory adverbial clauses, conditional (If ... , then ...) statements, and the like. A number of factors are at work here. As an (oral) storyteller proceeds along, each thing articulated quickly fades away as the next is uttered. But using complex sentences expressing relations between two clauses or sentences involves the presence (in some sense) not only of both of them but also of the relation. It is a more sophisticated activity than following a storyline one step at a time. This does not mean that members of non-literate people cannot think and speak in such patterns, only that writing encourages and reinforces them. The permanence conferred upon linguistic expression by writing makes it possible for all parts of complex locutions literally to be present at once. It thus makes such linguistic structures more familiar to literate persons and more likely to be used by them.

Possessed of permanent records and freed of the pressing need to preserve vital information as well as to adhere to cultural patterns that support its mental retention, members of literate societies can reflect on, analyze, and even question ancient tidbits of wisdom, as Robert Frost does regarding the maxim "Good fences make good neighbors" in his poem "Mending Wall." Similarly, when an action or a claim being urged, together with whatever support is being offered for it, is set down in writing, thoughtful assessment of the adequacy of that support is facilitated and promoted. Thus, the invention of writing paved the way for logic (which focuses on the structure of verbal expression) to emerge as something distinct from rhetoric (the effective and persuasive use of language). Approaching geometrical problems in this way led to the deductive proofs of the theorems that Euclid pulled together. A final point that needs to be made concerns the writing process itself. Writing does not merely provide a way to record information or the writer's thoughts; it involves a complex interaction between the writer and the written material. Once the process is underway, what has been written acts as a stimulus for reflection and for the exploration of avenues that were not anticipated. Writing on philosophical topics is notorious for leading people to the realization that they were mistaken in believing they understood something and had well-worked out views regarding it.

A final note of caution regarding orality and literacy is in order: the situation is far more complex, subtle, and fluid than can be conveyed in a short introduction. Scientists and scholars continue to refine, revise, and debate the issues involved. Like all introductions, this one should be viewed as a point of departure rather than a terminus.

The example of record keeping and what evolved from it illustrates extremely well the difficulties that must be overcome in order to understand how major technological systems come about and the significance of their invention. Clear parallels can be observed with the later differences in the development and spread of printing. The development of **printing**, like writing, occurred in China, but didn't have quite the same impacts as in the West. Certainly, it led to the spread of literacy and knowledge in both cases, but the vast industry that sprang up in Europe in the fifteenth century led to the exponential growth of available printed material. The development of movable type was less impactful in the logographic Chinese language, where a stock of thousands of characters might be required to print a text, rather than the far fewer individual letters of an alphabet that can be recombined. This meant that it was difficult for small printers and publishers to get started using movable type, and many continued using traditional woodblock printing, which worked well for their purposes. By contrast, publishing was widely bolstered by movable type in the West, where it was much more efficient and easier to implement even for smaller publishers, leading to more widespread adoption.

Also clearly demonstrated is the necessity of avoiding the assumption that technologies currently performing a certain function were invented to do so, presumably as the result of an inspired insight. But the most likely long-term benefit to be gained from this chapter is a better sense of the nature of number systems, including the recognition of their genesis as information technologies, and the impact of literacy. A point not likely to have been considered by most literate persons is that the transition from orality to literacy, like other technological revolutions, involved losses as well as gains.

One criticism of writing is that it undermines memory. Plato (in the *Phaedrus* 274a–277e) has Socrates raising the concern that people, relieved of the necessity of relying on their memories, will not make the effort to commit to memory things they really need to know and that they will mistake the ability to retrieve information from written documents for actually knowing. Of even greater concern to Socrates (and Plato, by proxy) is that static written information is more open to misinterpretation without the context provided by a present speaker who can elaborate or clarify intended meaning. Even the Sumerian record-keeping system could be indicted for its essential role in the oppressive, hierarchical political systems of Ancient Mesopotamia. Issues related to technology and society are addressed in Chapter VIII, those related to technology and the environment in Chapter IX.

Sources / Further Reading

Ascher, Marcia. *Ethnomathematics: A Multicultural View of Mathematical Ideas.* Brooks/Cole Publishing Company, 1991.

Bauman, Gerd, editor. *The Written Word: Literacy in Transition.* Clarendon Press, 1986.

Boyer, Carl B. *A History of Mathematics.* John Wiley & Sons, Inc., 1968.

Carlson, Keith Thor, et al., editors. *Orality and Literacy: Reflections across Disciplines.* University of Toronto Press, 2011.

Carter, Thomas Francis. *The Invention of Printing in China and Its Spread Westward*, 2nd ed. Revised by L. Carrington Goodrich. New York: Ronald Press, 1955.

Damerow, Peter. *Abstraction and Representation: Essays on the Cultural Evolution of Thinking*, vol. 175. *Boston Studies in the Philosophy of Science.* Dordrecht: Kluwer Academic Publishers, 1996.

DeFrancis, John. *Visible Speech: The Diverse Oneness of Writing Systems.* University of Hawaii Press, 1989.

Finnegan, Ruth. *Literacy and Orality: Studies in the Technology of Communication.* Blackwell, 1988.

Friberg, Jöran. "Numbers and Measures in the Earliest Written Records." *Scientific American*, vol. 250, no. 2, 1984, pp. 110–18.

Gaur, Albertine. *A History of Writing.* Charles Scribner's Sons, 1986.

Gelb, I.J. *A Study of Writing.* Revised ed. University of Chicago Press, 1963.

Glassner, Jean-Jacques. *The Invention of Cuneiform: Writing in Sumer.* Johns Hopkins University Press, 2003.

Goody, Jack. *The Domestication of the Savage Mind.* Cambridge University Press, 1977.

——. *The Interface Between the Written and the Oral.* Cambridge University Press, 1987.

Havelock, Eric A. *A Preface to Plato.* Cambridge University Press, 1963.

——. "The Linguistic Task of the Presocratics. Part One: Ionian Science in Search of an Abstract Vocabulary." *Language and Thought in Early Greek Philosophy*, edited by K. Robb. Hegeler Institute Monist Library of Philosophy, 1983, pp. 7–82.

——. *The Muse Learns to Write: Reflections on Orality and Literacy from Antiquity to the Present.* Yale University Press, 1986.

Huff, Toby E. *The Rise of Early Modern Science: Islam, China, and the West.* Cambridge University Press, 2017.

Ifrah, Georges. *The Universal History of Numbers: From Prehistory to the Invention of the Computer*, translated from the French. John Wiley & Sons, 2000.

Lieberman, Stephen J. "Of Clay Pebbles, Hollow Clay Balls, and Writing." *American Journal of Anthropology*, vol. 84, no. 3, July 1980, pp. 339–58.

Lord, Albert B. *The Singer of Tales.* Harvard Studies in Comparative Literature, vol. 24. 1960.

Luriia, Aleksandr Romanovich. *Cognitive Development: Its Cultural and Social Foundations*, edited by Michael Cole, translated by Martin Lopez-Morillas and Lynn Solotaroff. Harvard University Press, 1976.

Martzloff, Jean-Claude. *A History of Chinese Mathematics*. Springer, 1997.

Menninger, Karl. *Number Words and Number Symbols: A Cultural History of Numbers*, translated by Paul Broneer. MIT Press, 1969.

Nissen, Hans J. et al. *Archaic Bookkeeping: Writing and Techniques of Economic Administration in the Ancient Near East*, translated by Paul Larsen. University of Chicago Press, 1993.

Ong, Walter J. *Orality and Literacy: The Technologizing of the Word*. Methuen, 1982.

——. *The Presence of the Word*. University of Minnesota Press, 1983.

——. "Writing Is a Technology that Restructures Thought." *The Written Word: Literacy in Transition*, edited by Gerd Bauman. Clarendon Press, 1986.

Parry, Milman. *The Making of Homeric Verse. The Collected Papers of Milman Parry*, edited by Adam Parry. Clarendon Press, 1971.

Plato. *Phaedrus*. Reference numbers given are "Stephanus numbers," which refer to the pages of a standard edition that is used almost universally to identify specific passages.

Robins, Robert Henry. *A Shirt History of Linguistics*, 4th ed. Addison Wesley Longman, 1997.

Robson, Eleanor, and Jacqueline Stedall, editors. *The Oxford Handbook of the History of Mathematics*. Oxford University Press, 2009.

Schmandt-Besserat, Denise. *Before Writing. Vol. I: From Counting to Cuneiform*. University of Texas Press, 1992.

van der Waerden, B.L. *Science Awakening, Vol. I*, 4th ed., translated by Arnold Dresden. Leyden: Noordhoff International Publishing, 1975.

Vernant, Jena-Pierre. *The Origins of Greek Thought*, translated from the French. Cornell University Press, 1982.

Wolf, Maryanne. *Proust and the Squid: The Story and Science of the Reading Brain*. HarperCollins, 2007.

Zimansky, Paul. "Review of Schmandt-Besserat 1992." *Journal of Field Archaeology*, vol. 20, 1993, pp. 513–17.

Part Three:
Technology and the Modern World

Chapter VII:
Technology, Economics, and Science

SINCE IT IS THE very nature of technologies to provide ways of accomplishing things that would be more laborious, more difficult, or impossible without them, it is inevitable that the kind of economy that people have (how they satisfy their material needs and desires) is closely related to the technologies at their disposal. It is to be noted that such is the case with foragers as well as with agriculturalists, as was amply illustrated by the case studies in Chapters IV and V. Only because of their extensive technological repertoire were the Native Americans of the Pacific Northwest able to live as they did—or even survive—in that environment. Technologies not only enable people to occupy niches in the natural environment unavailable otherwise, they also foster conditions under which the human population increases dramatically, as happened in the case of agriculture. With farming comes sedentism, which fosters further technological innovation. People settled in different locations have access to different natural resources and (to various degrees) develop different skills and techniques. Under favorable conditions, this leads, naturally enough, to trade between groups of people and the production of goods specifically for that purpose, as seen in Chapter IV. Trade, in turn, tends to foster specialization, particularly if it takes years of practice to acquire the skills needed to make a product for which there is much demand, as exemplified by canoe-making in the Pacific Northwest.

The extent to which **commerce** is carried on depends upon the availability of transportation and the desirability of the products exchanged, again as illustrated by the waterborne trade carried out in the Pacific Northwest and the willingness of the Native Americans to pack fish oil (on their backs) over the coastal mountains via the "grease trail" into the interior to trade for animal skins. The location of so many of the world's great cities adjacent to rivers or harbors testifies to the importance of waterborne transport. But there would have been no large cities had there not been ways of transporting large quantities of agricultural products over land (just to get them to bodies of water) that were more efficient than carrying them on the backs (or heads) of human beings. Such tasks were accomplished by harnessing the muscle power of animals to carry loads and to pull carts, wagons, sleds, etc. As noted in Chapter V, a very important aspect of the domestication of cattle was the development of a source of **motive power.** Since dogs are believed to be the first animals to have been domesticated, they very well may have been the first draft animals. Although horses probably epitomize animal-powered transportation to most people, they were by no means the first to supply it.

Of course, even waterborne transport requires motive power, the earliest of which probably was provided by means of poles or paddles. Subsequently, at various times and places, people devised ways of capturing **wind power** for this purpose. Eventually they developed one configuration or another of these "sails," as they are called, that were very effective, enough to make transoceanic travel possible. As will be discussed in Chapter VIII, the Chinese sent out several impressive, long-distance marine expeditions in the early fifteenth century, reaching the east coast of Africa before withdrawing into their own coastal waters. Other ways of utilizing wind power were devised at other times and places, perhaps most notably for pumping water and milling grains. Recently, the effort to exploit "renewable" sources of energy has resulted in the design and construction of very large windmills. Beginning during the same century that Chinese marine activities peaked, Western European nations launched numerous voyages of exploration, commercialization, and colonization that spanned the planet. Such effective exploitation of forces other than muscle power, whether human or animal, is a recent phenomenon when viewed against the background of human history and prehistory. Indeed, the extensive development of such sources of motive power, the power to do work, is a fundamental feature of the modern world, distinguishing it, perhaps more than anything else, from other times and places. The first case study of this chapter, "From Motive Power to Energy," reviews some of the key steps taken, beginning in Roman times, not only in harnessing one or another of the powers of nature but also in developing technologies that these sources of power made possible.

Viewed retrospectively, the acquisition of sources of motive power is but a part of a much larger, ongoing process encompassing areas of human activity once not thought to be related. These have been brought together under the same rubric as the result of advances in the basic sciences which demonstrate that a number of seemingly unconnected phenomena are actually manifestations of the same underlying principle, according to which something called "energy" can assume many forms but cannot be created or destroyed. This development is a close parallel of an example cited in Chapter II: electrical phenomena being recognized as a class because they were all shown to be implications of the same set of laws, laws worked out piecemeal over an extended period. Similarly, the principle of the conservation of energy emerged in a series of steps in which it was established empirically that certain equivalences hold, perhaps most notably that between heat and mechanical energy. Prior to the eighteenth century, hardly anyone would have thought that heat, which was closely associated with fire, was in any way connected with motive power; nor could anyone at that time even have imagined that both heat and mechanical energy might be transformed into electrical energy (very little then being known about electricity). Nowadays in industrialized parts of the world, access to energy is central to the lives of individuals as well as to the economies of nations, as becomes evident when there is an interruption of electrical service, a "brown out," or a shortage of gasoline or natural gas.

The encompassing nature of the energy conservation law was not recognized until the nineteenth century, and only in the latter part of that century did science, specifically chemistry and physics, assume a major indispensable role in technology, providing foundations for the chemical and the electrical and (electronic) communications industries, as noted in Chapter II. These developments mark the formation of a powerful partnership, usually referred to as "science and technology," even though "technology and science" would seem as appropriate. The favored usage reflects (i) the prevalent view that science is the more basic of the two enterprises and (ii) the maxim, formerly oft-repeated, that technology (properly understood and practiced) is "applied science." It also embodies (iii) the recognition that the modern engineer is a highly educated professional, rather than an artisan trained in an apprenticeship, and that study of the basic sciences is a key part of that education. Because of the dependence of modern science upon its technological repertoire and the inevitable economic ties resulting therefrom, economic considerations could hardly fail to influence the course that scientific inquiry takes. This topic is addressed below in section B. Examples discussed include the development of quadrupole gas chromatography/mass spectroscopy and laser cooling and trapping.

A. From Motive Power to Energy

1. Water Power

Despite their engineering prowess, exhibited in the building of roads, aqueducts, bridges, arches, barrel vaults, domes, etc., the Romans contributed very little toward harnessing the motive powers of nature to do useful work. There are reasons to believe that their merchant ships were able to sail at an angle into the wind and thus to "tack" (zigzag) in that direction, even if their method of accomplishing this is uncertain. In another important development, **water-powered grain mills** appeared by 100 BCE in northern Greece (then Roman territory). Some three centuries later, the Romans constructed a very large, grain-mill complex at Barbegal, near present-day Arles, France (see Figures 7-1 and 7-2 below). Unlike the waterwheels of the earlier mills, which were "**undershot**" (powered by flowing water pushing against the bottom paddles of the wheel), those at Barbegal were "**overshot**" (powered by water flowing from a chute onto the top of each waterwheel). In an overshot waterwheel, water fills each scoop-like pocket of the wheel when it is at the top; this water is dumped out at the bottom in each revolution. It is primarily the weight of the water in the full pockets on the descending side of the wheel that powers it. In an efficiently designed facility, only a small fraction of the torque (turning power) of the waterwheel derives from the actual motion of the water flowing onto it. Although the overshot configuration is much more efficient and more powerful than the undershot, it requires a much more elaborate (and expensive) installation: a reservoir of water (typically necessitating the construction of a dam) and a flume or millrace to carry the water from it to the top of the millwheel. The sixteen overshot waterwheels at Barbegal were arranged in pairs in a line extending down a slope from a reservoir which was fed by an aqueduct. Thus, water released from the uppermost waterwheel was directed to the top of the wheel just below it, and so on. Each waterwheel turned two millstones by means of a system of gears. Although undershot watermills are simpler and more inexpensive to set up, they suffer from the disadvantage that their performance is vulnerable to fluctuations in the flow of the water supplied to them: too little translates into loss of power, while high water can render them inoperable. An effective method of addressing this problem, widely employed

Figure 7-1. The Ruins of Barbegal. This photograph shows the current state of the Roman mill complex at Barbegal. Image credit: MOSSOT. CC BY SA 4.0.

during the Middle Ages, is to mount these mills on barges anchored in rivers. Indeed, these floating mills became so numerous in certain favored locations that they were an obstacle to river navigation.

The Barbegal complex is such an impressive achievement that one might have expected the Romans to have exploited this new power source for other purposes, but such applications appear to have been quite limited. There seems to have been very little incentive to develop labor-saving devices and techniques, due to (i) the availability of slave labor, (ii) the existence of significant unemployment prior to the decline of Roman power, and (iii) an administrative structure geared to discharging specific responsibilities in recognized ways. In spite of this general

Figure 7-2. Model of the Roman Mill Complex at Barbegal. This model at the Museum of Ancient Arles represents the ancient Roman mill complex at Barbegal near modern-day Arles, France. Image credit: Carole Raddato. CC BY SA 2.0.

pattern, there are reasons to believe that the Romans utilized water power to operate mills of a second type: **sawmills** for cutting marble. These mills embodied a **crank-and-connecting-rod mechanism** that makes possible the conversion of rotary motion into back-and-forth (or up-and-down) motion and vice-versa. This basic concept is so important—it is fundamental to both steam and internal combustion engines and to countless other devices—that a brief explanation of it is in order.

A simple crank is little more than a rod or shaft with two 90° bends, giving it a shape much like that of a "Z." If one end of it is affixed to the center of a wheel or similar object (mounted so that it can rotate), moving the other end of the rod in a circle will cause the wheel to rotate. Such an action is referred to as "cranking." A good example is the front wheel of a child's tricycle, which is turned by two cranks (one for each foot). The axel serves as one of the end-sections of both cranks. A pedal makes up the other end-section of each of the cranks. The center section of each crank extends from the center of the wheel out to the pedal. Moving the pedals in a circle ("pedaling") causes the wheel to rotate. Another way to achieve the same result is simply to attach a short rod or handle perpendicular to the face of the wheel near its perimeter, as is found in an eggbeater. Then the wheel can be made to rotate (can be cranked) by means of this handle.

The connecting rod is a device that makes it possible to use oscillatory motion to produce circular motion or vice-versa. Usually it takes the form of an oblong piece of material (typically metal) with one end attached to a crank and the other end to something that moves only up-and-down or back-and-forth. Early in the Han Dynasty (207 BCE to 220 CE), the Chinese invented a hand-crank-powered machine to separate rice grains from the husks and stalks. Subsequently, they added a foot pedal connected to the machine's crank by a rod so that it could be powered by the up-and-down motion of the operator's foot, leaving the hands free for other tasks. In more recent times in the West, sewing machines were powered in this way until electricity and small electric motors became available. Similarly, the reciprocating motion of pistons linked to crankshafts, in the case of automobiles, or to short cylindrical shafts projecting from the sides of wheels, in the case of steam engines, causes their wheels to turn. Such mechanical systems can, of course, be operated in the reverse order: an object can be made to move in an oscillatory manner if it is linked to a rotating wheel by a crank-and-connecting-rod mechanism. It seems that this is how the Roman sawmills worked: frame saws were pulled back-and-forth by connecting rods attached to short cylindrical shafts (cranks) projecting from the sides of waterwheels. When the cattle farmers moved into the relatively dry grasslands of North America, they had to find a way to supply water to their herds and to themselves. The solution involved utilizing cranks and connecting rods to produce up-and-down motion (needed to pump water) from the rotary motion of windmills.

However limited may have been the kinds of uses the Romans made of water power, grain mills thus powered could be found throughout what is now France and the Low Countries before Roman rule ended and across Europe by the end of the first millennium. When, following their conquest of England in 1066, the Normans completed their survey of the population and resources of the land (recorded in the "Domesday Book"), it was found that there were more that 5,600 water-powered mills south of the river Trent. Undershot millwheels were employed where streams flowed evenly along relatively flat courses, and overshot ones were used where streams flowing swiftly downhill could be dammed and millraces constructed to carry water to the top of waterwheels.

Since a dependable supply of water is vital to water-powered industry, it can hardly be surprising that the next major step in the development of water-powered machinery took place on the north side of the Alps where there is rainfall throughout the year, in contrast to the Mediterranean climate (winter rains and summer drought) to the south. But this advance did not take place right away, nor did it involve the crank-and-connecting-rod mechanism. Once the Western Roman Empire had passed into history and the quarrying of marble was suspended, such mechanisms would not reappear in the region until the Late Middle Ages. In the ninth century simple cranks were used to turn grindstones by hand, but no industrial applications were developed until much later. Instead of mechanisms involving cranks, the workhorse of medieval industry was the **cam**, another device that converts rotary motion into reciprocating (up-and-down or back-and-forth) motion. A cam is little more than a protrusion, a "bump," on the side of a rotating shaft. In the simplest arrangement, the protrusion pushes against an object each time the shaft completes a rotation. If the object returns to its original position after the bump passes (for example, because of its weight or more recently by the action of a spring), the result is an up-and-down (or back-and-forth) motion. Like crank-and-connecting-rod mechanisms, cams play an important role in internal combustion engines: opening and closing valves. By arranging for the cam to depress one end of a lever secured by a pivot, the up-and-down motion of the other end can be applied to tasks that involve pounding or hammering.

The Chinese had used such devices, called "**trip-hammers**," as early as the third century CE to break the husks loose from grains of rice, to pulverize ores, and to forge metal objects. And although the Ancient Greek engineers were acquainted with the concept, they exploited it only in mechanical toys and gadgets. It remains an open question whether or not those who initiated the use of cams in Medieval Europe knew of these earlier applications. The earliest recorded use (890 CE) of a water-powered, cam-activated hammer in Europe was to crush malt in a beer-making operation at a Benedictine monastery in Switzerland. Many other uses were found for the device during the four or so centuries that followed, a period that some have called the "Medieval Industrial Revolution." As noted in Chapter IV, obtaining fiber for linen cloth involves beating rotted flax stems to loosen the decomposing soft parts from the fibers. By 990 CE this work was being done by water power. In the following century, water wheels were harnessed to trip-hammers to forge iron.

Perhaps the most economically important early such use was in **fulling** wool cloth. Once woolen thread has been woven into cloth, it must be "fulled," soaked in water with cleaning agents and pounded to remove grease and to soften it and make it bulkier. In Roman times, workers accomplished this by putting the cloth in a water-filled trough and tramping up and down on it with their bare feet. Spinning thread and weaving it into cloth was still done by women at home; but finishing operations like fulling and dying was done in urban shops. Not by accident was the production of wool and linen cloth a major element of the Roman economy. Except under special circumstances, as in the case of the Magdalenians (noted in Chapter IV), or when the human population was low enough that the supply of animal skins sufficed, human beings were able to live outside the warm equatorial belt only if they had devised a way to produce some kind of fabric to protect themselves from the elements. As seen in Chapter IV, the Native Americans of the Pacific Northwest exhibited great ingenuity in meeting this challenge.

An important factor heavily influencing the extent to which the powerful overshot waterwheel was used and new applications developed was the investment required for such installations, as noted above. Furthermore, since new mechanical devices take time to make and remake until they function in the desired way, the impressive variety of water-powered devices that appeared during the Medieval period would never have seen the light of day had there not been an enduring social unit or organization with the resources and commitment to support such endeavors on a long-term basis. Given the conditions then prevailing in Western Europe, it is not surprising that the crucial organizations were religious in nature. What may be surprising is that monasteries provided the economic support necessary for the development of so many of the technologies that, in turn, transformed the economy of Western Europe.

This state of affairs has its roots in the founding of a monastic order, the Benedictines, at a time (shortly after the dissolution of the Western Roman Empire) when such a community would have to be self-sufficient or else perish. The key to the survival—and the prosperity—of the order was the recognition of work as one of its core values, on a par with prayer and study. This depth of commitment is expressed by the maxim "*Laborare est orare*" (To work is to pray). Each monastery operated as an independent, self-supporting organization, presided over by an abbot. Membership was understood to replace having a family, and brothers were expected to devote themselves completely to the monastic life; they took vows of poverty, chastity, and obedience. Over the centuries, lots of hard labor and adherence to the rules set down by St. Benedict, the founder of the order, led to accumulated wealth. So much so that by around 1100 CE the contrast between the opulence of the monasteries and St. Benedict's vision of the monastic life prompted some reformers to strike out on their own. In order to separate themselves from worldly enticements and to ensure that hard physical labor would be essential to the survival of the community, the Cistercians, as they are called, selected sites for their monasteries in remote locations on unproductive land, such as the upper reaches of mountain valleys. Of necessity, they quickly became expert at establishing productive operations on marginal land. They not only worked hard, they gathered valuable information, honed their management skills, and used their ingenuity to develop new techniques and devices to assist them in accomplishing their tasks. Most significantly, there was an awareness among them not only of the benefits of implementing existing technologies but also of the possible benefits of putting existing technologies to new uses. Doing so was not only permitted but encouraged by the leadership of the order.

Another important way that the monasteries and other religious institutions contributed to the genesis and development of technological innovation during the Medieval era was by providing an effective communication system. For centuries following the fall of Rome their network was the primary, if not the only, means by which not only messages but also books and, in some cases, other goods could be sent from one place to another. Sharing information and techniques, whether by means of missives or meetings, played a vital role in the flowering of technology in Medieval Europe. During this period, monasteries provided yet another valuable service: copying books and other documents to replace worn-out ones and to supply the needs of new monasteries, churches, and other organizations. Since it had to be done by hand, copying was a slow and laborious process. It was not confined to religious material but included a wide variety of writings from the ancient world, among them scientific and technical ones.

The Cistercian order grew rapidly, as did the number and the types of water-powered factories and other enterprises that they inaugurated or operated. Within slightly more than a century of its founding, the order had expanded to more than 500 houses. In the twelfth century there were 750 of them. Each of the Cistercian monasteries was a self-contained factory community, featuring (in addition to the church and living quarters for the monks) a number of shops and **water-powered industrial operations.** They were located on or near streams, not only to secure a source of power but also to provide for domestic needs and to carry away wastes. Although the types of industries engaged in varied, depending upon local conditions and the availability of raw materials, certain operations could be found at essentially all of the abbeys: milling grain and processing fiber and cloth. Many of the monasteries also established vineyards and produced wine, employing water-powered winepresses in the process. Where olives grew, water-powered machinery was used to extract olive oil. Where iron ore could be obtained, trip hammers were used to crush the ore and to forge the **iron** being worked.

2. Water Power and Iron

Until powerful bellows were developed, it was not possible to raise the temperature of iron high enough to melt it. It was (and is) possible, however, to refine iron ore and work iron at a lower—but still very high—temperature. If pulverized iron ore (which is an oxide or similar compound) is heated in an intensely hot charcoal or coke fire (see below), the carbon will combine with the oxygen and other elements in the ore forming gases which escape, leaving a sponge-like mass of porous iron. Then, because iron softens at temperatures then obtainable (about 1450°F), it is possible to pound the spongy mass into a solid block of "bloomery iron," as it is called. But achieving this requires repeatedly reheating and pounding the metal. This process, referred to as "**forging**," also removes impurities; but it is very hard work. In keeping with their usual approach to such challenges, the Cistercians put water-powered trip-hammers to work crushing the ore and forging the metal. They are believed to have operated water-powered forges in France at an early date and were the first to do so in Germany, Denmark, England, and southern Italy.

Much, if not most, of the technological and economic significance of iron derives from the **alloys** (homogeneous mixtures) that it forms in combination with other elements, most notably carbon. **Steel**, a very tough and important material, is an alloy of iron and carbon. In contrast, pure iron is soft and ductile, making it inferior to bronze for many purposes. If iron is heated for an extended period while in close contact with charcoal, it will absorb carbon in somewhat the same way that water dissolves salt. The amount of carbon absorbed, and thus the properties of the alloy formed, will depend on the temperature and the length of time that the iron is heated while in contact with the charcoal. Ancient metal smiths of the Eastern Mediterranean region appear to have initiated the practice of "**carburizing**" iron, as it is called, by around 1200 BCE. Of course, these ancient craftspeople had no way of knowing that part of the fuel they were using to heat the iron was actually being absorbed by it; nor could they have known that the desirable changes taking place in it were the result of the formation of an alloy. With charcoal being used to refine the iron ore and to heat the metal repeatedly as it was worked, inevitably a certain amount of carbon would have been absorbed. And, under the conditions then prevailing, there was bound to have been some variation in the characteristics of the final products. Presumably people noticed the superior qualities of some of them and sought to produce more of the desirable types. Somehow—whether by accident, trial-and-error, educated guess, or some other means—ancient iron workers learned to increase the ratio of charcoal-to-iron ore in their furnace charges, to intersperse layers of the ore with layers of charcoal, to extend the time that the ore was heated, to raise the temperature achieved, and thus, unbeknownst to them, to increase the carbon absorbed. They learned to make steel. There could hardly be a better illustration of technological innovation proceeding unassisted by scientific understanding. As noted at the beginning of the chapter, however, and as will be discussed fully below, science is an important resource for technology; and both have flourished in the last couple of centuries in a powerful partnership.

The method that was developed to raise the temperature at which iron ore was smelted is especially important. As has been known and exploited since as early as *Homo erectus* times, combustion can be accelerated (a fire can be made to burn hotter) by directing an air current at the burning fuel. For this purpose, people have utilized quite a variety of devices, from blowpipes to bellows. A breakthrough in iron production and processing was achieved when **waterwheels** were put to work operating the bellows that forced air into the furnaces in which iron ore was being smelted. This use of waterpower makes it possible to achieve a temperature that is high enough for the iron in the ore to absorb enough carbon (about 4 percent) for an alloy to be formed that melts at a lower temperature (2100°F) than does pure iron (2800°F). This innovation also made it possible to raise the temperature of this iron/carbon alloy to its melting point. The molten alloy can then be poured into molds, as had been done much earlier with bronze (which melts at about 1750°F). Articles produced in this way are said to be made of "**cast iron.**" The production of cast iron tools and weapons was an established part of Chinese material culture by the fourth century BCE. But it was only in the fourteenth century CE that Europeans first succeeded in liquefying iron by the method described above. Although borrowing from the Chinese is a possibility, differences between the two systems raise doubts regarding such an influence. The European approach featured an overshot (vertical) waterwheel that powered the bellows by means of cams; whereas the Chinese employed a horizontal waterwheel (which functions very much as an undershot vertical one) that operated a very differently structured bellows by means of a crank-and-connection-rod system.

Just as the ancient metal smiths learned (somehow) to carburize iron to make it harder, so ironworkers had learned, by the beginning of the modern period, to include limestone in the charges they put in their furnaces. Although they had no way of knowing it, the limestone ($CaCO_3$) reacts chemically with the impurities in the iron ore to form compounds (collectively referred to as "**slag**") that liquefy at a lower temperature than do the impurities in their original state. As a result, once the iron and the slag are in a molten state, they form separate layers in the furnace because of their different densities (just like oil and vinegar in a bottle). Then, depending upon the structure of the furnace and the tools and techniques available, the slag can be skimmed off or drained through a hole in the wall of the furnace (which would have been kept plugged with fire clay until needed), leaving the heavier molten iron at the bottom of the furnace to be drained or dumped as well.

This method of smelting iron ore has tremendous advantages. It yields an iron alloy that is relatively free of impurities and is in molten form so that it can be cast in essentially any desired shape. But cast iron is of limited usefulness because of its brittleness. (Perhaps it should be noted at this point that the percentage of carbon in steel, about 0.35 percent for most steel, is much lower than the percentage of it in cast iron, up to 4 percent.) Somehow, the European ironworkers learned to address this problem effectively by adding a second step to the smelting process. First, they let the molten iron flow from the furnace into a system of shallow trenches, there to cool and solidify. Because of the resemblance of these chunks of iron to a litter of pigs suckling their mother, this product came to be referred to as "**pig iron.**" Next, these "pigs" were reheated in a second furnace while being subjected to a blast of air from water-powered bellows. Again, the early ironworkers would have had no way of knowing it, but the effect of treating pig iron in this way is to remove the carbon in it. The oxygen in the air blast combines with the carbon in the pig iron to produce gases that escape, leaving "**wrought iron,**" as it is called, which contains very little carbon and a small amount of residual slag. It is ductile, rather tough, but soft enough to be cut by steel chisels, saws, and drills.

Subsequently, the chunks of wrought iron were pounded by water-powered trip hammers and intermittently reheated (utilizing water-powered bellows) as they were formed into pieces of iron stock (typically bars of rectangular cross-section) to be supplied to blacksmiths. At some point, water-power was put to work turning pairs of iron rollers which squeezed pieces of wrought iron into flat strips for such purposes as making nails. Operations of the kind just described were carried out at the first iron smelting, refining, and processing complexes constructed in North America. The earliest successful such venture began production in 1647 at a site on the Saugus River in Massachusetts.

3. Coal and Steam Power

It is only a small step from iron smelters of the type just described to full-blown **blast furnaces** that operate continuously. All that is required is for air be forced into the furnace near the lower end, the molten iron to be drained out at the bottom, the slag to be drawn off just above it (also through a hole in the side of the furnace), and iron ore, charcoal, and limestone to be added at the top—all continuously. There was only one more obstacle to be overcome in order for people to have access to quantities of iron previously not thought possible. What was missing was the enormous amount of charcoal that would be required. Charcoal was essential because higher temperatures can be achieved with it than can be produced with wood. Wood is used to make charcoal, and wood was in such short supply in England in the 1600s that coal was rapidly replacing it as a fuel. Coal, however, has some serious drawbacks; not the least of which is the dirty, black smoke that coal fires release. The sulfur in coal is especially problematic. Early attempts to use coal as a source of heat in making beer were thwarted by the effect of the coal smoke on the taste of the beer. But brewers found that subjecting coal to the same treatment that produces charcoal (when applied to wood) eliminates the noxious fumes: they baked the coal in airtight enclosures (depriving it of air) before using it as fuel. This process yields "coke," a spongy form of carbon very like charcoal. Similarly, initial attempts to use coal in smelting iron ore got nowhere; but during the first half of the eighteenth century, Abraham Darby (1678–1717) and his son successfully demonstrated the use of coke in smelting iron ore and in producing wrought iron from pig iron. The use of coke quickly spread throughout the industry and soon replaced charcoal as the primary fuel for such purposes.

Unfortunately, mining coal is plagued by problems intrinsic to the enterprise. Particularly vexing is water coming into and flooding the mines. As the demand for coal increased, this problem became more acute. With operators availing themselves of any means possible to remove the water, including horse-powered bucket-chains, enterprising people began looking for ways to address the problem by taking advantage of recent advances in human understanding of vacuums, gases (including steam), vaporization/condensation (including the vast difference between the volumes of the liquid and the gaseous state of a substance), and the effects of heat and pressure (or their absence). The first (at least partially) successful such undertaking utilized (i) steam pressure to push water up a pipe and (ii) the vacuum produced by cooling steam in a sealed container to draw water up from a pool below. This devise was designed and patented in 1698 by Thomas Savery (c. 1650–1715). Although it was found to be useful for pumping water for towns and large buildings, this pump proved to be impractical for draining mines.

Subsequently, it occurred to Thomas Newcomen (1663–1729) that a **piston** might be made to move up and down inside a hollow cylinder by the application of steam pressure (to push it up) followed by the condensation of the steam (induced by cooling) to allow it to come back down. (Pistons moving inside hollow cylinders were in common use for a variety of purposes at the time. See below.) Newcomen realized immediately that since up-and-down motion is precisely the kind that is involved in the operation of a lift pump, the desired performance of such a pump could be achieved by connecting the pump rod to an oscillating piston by means of a rocking beam. Right away, he set about constructing a pump thus powered. As often happens, the key insight that led to a successful pump and opened the door to the steam age was occasioned by an accident. Following Savery's lead, he enclosed the cylinder of an early model of his device in a metal jacket into which cold water could be introduced to cause the steam in the cylinder to condense. The cylinder he used had a casting fault that had been repaired with tin solder. In a test run, the intensely hot steam melted the solder, permitting the cold water to rush into the cylinder, instantly condensing the steam and creating a high vacuum. The unbalanced atmospheric pressure on the top of the piston exerted such a force on it that the piston knocked the bottom out of the cylinder and smashed the top of the boiler beneath it.

Recognizing the significance of what had happened for his project, Newcomen immediately set to work modifying his mechanism to take advantage of the phenomenon (see Figure 7-3 below). The basic structure of his device remained the same. The piston was connected (from above) to one end of a rocking beam, from the other end of which hung the pump rod. As planned, the upward motion of the piston resulted from steam being introduced into the cylinder below the piston. The change that Newcomen made in his original design was to arrange for cold water to be sprayed into the steam-filled cylinder when the piston reached the top of its stroke. The resulting condensation of the steam left a high vacuum so that the unbalanced atmospheric pressure on the top of the piston pushed it back down, which raised the other end of the rocking beam, operating the pump. Because of this utilization of atmospheric pressure (to do the work of lifting water), Newcomen's invention (1712) is called an "atmospheric engine" rather than a "steam engine." Also called a "beam-engine," it quickly demonstrated its functionality and cost effectiveness and was not only put to work across Britain but was exported as well. An initial barrier to its widespread adaptation, the cost of the brass cylinders, was overcome when it was established that the cylinders could be cast from molten iron at a much lower cost.

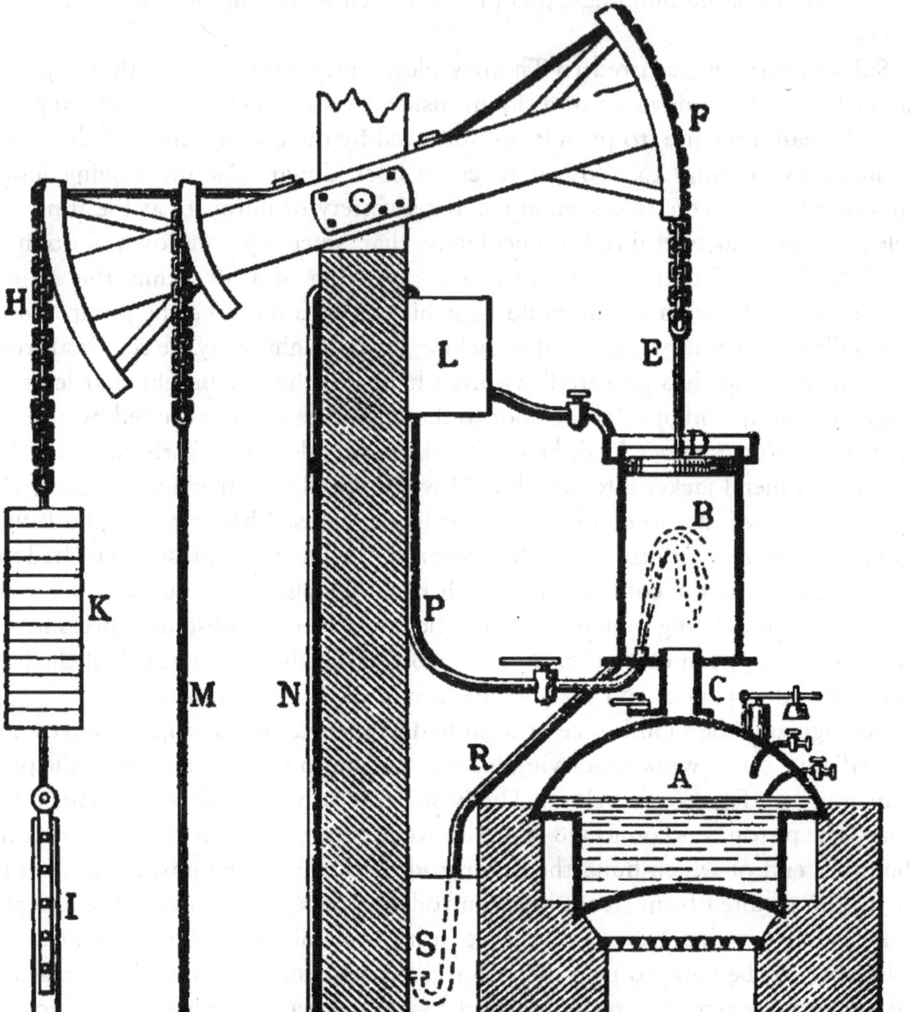

Although Newcomen's engine actually was very inefficient (see below) and limited in purpose and capabilities, it brought together three elements that were to play intertwined, central roles in the Industrial Revolution: *iron*, *coal*, and *steam*. Among historians and anthropologists, the "Iron Age" is considered to have begun when iron technology first made its appearance some three thousand to thirty-five hundred years ago. But nowhere and at no time prior to

Figure 7-3 (opposite). The Newcomen Engine. The steam is produced in the boiler A and reaches the cylinder B through the pipe C. The steam in the cylinder forces the piston D upwards, as in Papin's invention, assisted by the counterweight K. At the piston D the force is coupled via the fixed rod E and in turn by a chain to the beam F. The counterweight K is attached via the chain H to the other side of the beam. The pump rod I is rigidly coupled to the base of the counterweight, being pushed up and down by the corresponding movements of the counterweight. By stopping the flow of steam through pipe C, by use of the stopcock, the contents of the cylinder are closed off. Water from the reservoir L is injected into the cylinder via the pipe P. (This point in the cycle is shown in the illustration.) The injected water speeds up condensation of the steam in the cylinder, causing a vacuum. The piston is pushed downward by the air pressure from outside, moving the mechanically coupled beam F as well. This movement lifts the counterweight K and the coupled pump rod I upwards. The pipe R takes away the condensed water, S marks the section of the U-shaped pipe submerged in the reservoir, closing off the cylinder in a steam tight manner. M is the linkage of a small auxiliary pump which fills the reservoir L through the pipe N. In the original implementation, the stopcocks were operated manually. Adapted from the 4th edition of *Meyers Konversations-Lexikon*, 1890.

1700 did iron technology dominate material culture the way it did in industrialized countries by the 1850s. This point is graphically illustrated by the contrast between the wooden gears of earlier industrial installations (from those of the Roman mills at Barbegal to those of the Saugus iron works in Massachusetts) and the complex, mostly-iron machines stretching across nineteenth-century factory floors. The case of coal is similar. Initially it was mined simply because, like wood, it could be burned to produce heat for the usual purposes: keeping warm, cooking, and the like. But once people learned to make coke from it and had steam engines fueled by it, coal assumed a central role in their economic lives. Not only did Newcomen's engine or pump open the door to large-scale coal production, it also burned coal to produce steam, as did the thousands of steam engines that came after it.

The shift from wood to coal (and from charcoal to coke) as a source of heat marks a watershed in the development of Western technology: from a reliance on wood, animal and vegetable fats and oils, dried animal droppings, and the like for heat and light to the use of fossil fuels such as coal, petroleum, and natural gas. The abundance of coal (and coke) made possible an enormous increase in the production of iron without which all those trains, rail lines, steamships, machines in factories, etc. could never have been built. As more iron became available, more uses were found for it and new capabilities for processing it were developed. The harnessing of steam to provide motive power represents a radical departure from previous methods (which relied upon muscles, water, wind, and, for certain limited purposes, such things as gunpowder) in that it involves the exploitation of a relationship between aspects of the world previously thought to be very different and independent: motion and heat—or, perhaps better, fire.

At the time, Newcomen's machines were called "fire engines" in recognition of what was seen as the basis of their operation: the **motive power of fire.** In his 1769 patent application for the improvements he made in these devices (see below), James Watt referred to them as "fire engines." Sadi Carnot (1796–1832) even used the title, *Reflections on the Motive Power of Fire*, for his pioneering work on thermodynamics, published in 1824. In it he demonstrated that there was a theoretical limit to the efficiency with which heat engines (including steam engines) could operate.

In the popular mind, the name of James Watt (1736–1819) is more closely associated with the invention of the steam engine than any other. And rightly so. However, to say that he "invented the steam engine" is not only a gross oversimplification but is inconsistent with a major theme of this work: by and large, major technological innovations take place in a series of steps, rarely if ever initiated to achieve their eventual capabilities, and often involving multiple people. As noted below, other scientists, engineers, and entrepreneurs such as Joseph Black, Matthew Boulton, and John Wilkerson all played a role in the development of the steam engine. Watt made a series of key improvements in the "fire engine," beginning with modifications to Newcomen's beam engine and culminating in a device that transformed many aspects of the human scene.

Watt's attention became focused on steam engines when, as a young instrument maker/repairman in Glasgow, he took on the assignment of repairing a model Newcomen engine being used for demonstration purposes at the university. Reflecting on how the device functioned led him to realize that there were fundamental inefficiencies in the design of the engine. He began to think about how steam was being wasted and how the machine might be improved. Contributing to Watt's thinking on these subjects was professor Joseph Black (1728–99), who befriended Watt at the University of Glasgow in the late 1750s. Black's discoveries concerning the nature of interactions of heat with matter influenced Watt's experimentation with the engine, further illustrating the powerful partnership of technology and science. Crucially, Watt noted that the cold water, sprayed into the cylinder to trigger the condensation of the steam, cooled the cylinder itself as well. Consequently, a significant fraction of the steam entering the cylinder during the following upstroke was wasted on reheating it. It occurred to Watt that if the steam-filled cylinder were connected to an evacuated container to which a valve opened at the end of each upstroke, the steam would rush from the cylinder into this auxiliary chamber and could be condensed there without the cylinder being cooled. The vacuum could be created by connecting the "**condenser**" (as this auxiliary container is called) to a vacuum pump. In order to test his idea, Watt proceeded to construct a small prototype. To facilitate the condensing process, he immersed in a cold-water bath both the condenser and the vacuum pump used to evacuate it. When the model was provided with steam from a small boiler, its performance met Watt's expectations. He added yet another feature to a larger model: a hot water/steam jacket surrounding the cylinder (see Figure 7-4 below).

Granted a patent in 1769, Watt lacked the means to launch a manufacturing operation himself. Eventually (in 1773), he had the good fortune to enter into a partnership with Matthew Boulton (1728–1809), a manufacturer of light metal products in Birmingham. Just as the Cistercians had provided the resources and commitment essential to the development of water power in the Middle Ages, so Boulton provided the wherewithal that brought Watt's inventions to fruition and opened the door to the development of steam power. Not only did his factory boast state-of-the-art equipment and highly skilled workers, but Boulton had experience coordinating the efforts of people of a variety of technical skills; he had the economic resources to see the project through to a successful conclusion and the commitment to do so. One of the challenges the partners faced was obtaining iron cylinders that were both as large and as precisely bored as they needed. Luckily for them, John Wilkinson (1728–1808) had just invented a precision boring machine for use in making cannons, which enabled him to supply cylinders that met their requirements. Initially, they undertook the construction of two full-sized steam engines incorporating Watt's ideas. As with all new complex systems, unanticipated difficulties were encountered, and it took time to "get the bugs out." But in 1776 one of the engines was put to work at a coal mine, pumping water, the other at a blast furnace, powering the bellows. The improvement in efficiency was nothing less than spectacular: fuel consumption was reduced to one third of what a Newcomen engine would have required to perform the same tasks. Following that, orders poured in, and Boulton's Soho factory at Birmingham was kept busy.

Watt pressed on with additional innovations. Three of these stand out: (i) a steam engine that delivered rotary motion (rather than up-and-down motion), making it a power source for all kinds of machinery; (ii) a **double-acting cylinder**, in which steam pressure is applied alternately, first to one side of the piston and then to the other; and (iii) a centrifugal **governor**, a feed-back mechanism that enables an engine to maintain a constant speed even under varying loads by reducing the flow of steam when a certain speed is exceeded and increasing

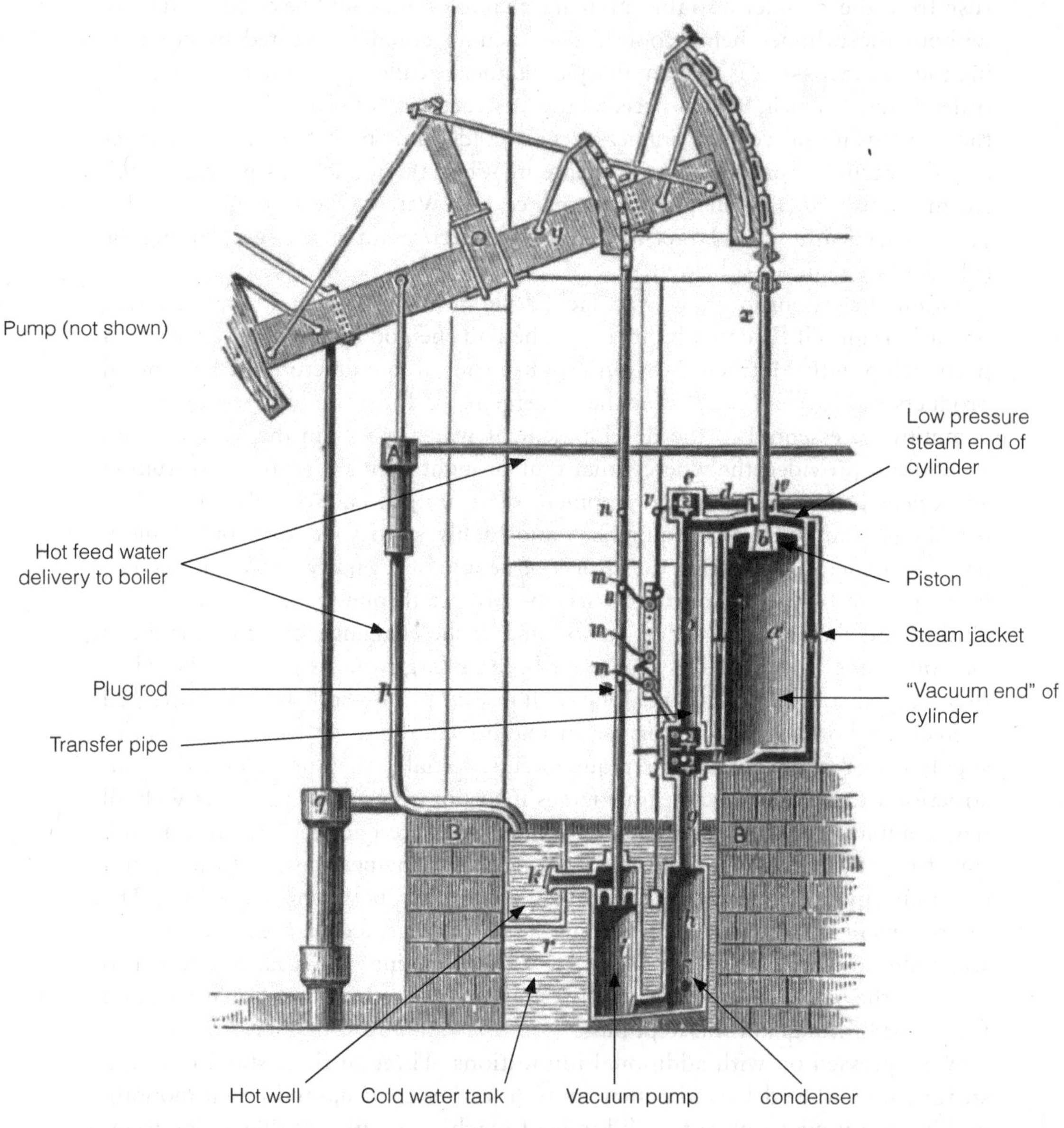

Figure 7-4. Watt Steam Engine. Note several of Watt's crucial additions to the Newcomen engine here, including the vacuum pump, condenser, and steam jacket. Adapted from Robert Henry Thurston's 1878 ***A History of the Growth of the Steam Engine***.

it when the speed falls below a certain level. These three new features were essential to the role steam engines were to play in the Industrial Revolution and thereafter; but they are not, to the same degree, inventive achievements. Of the three, the first and third involve little more than attaching or adapting an existing device to Watt's steam engine. But the second involves fundamental changes in the way the machine operates. By reconfiguring how the valves and conduits connect the boiler and the condenser to the cylinder, Watt arranged for steam to enter the cylinder on one side of the piston while the space on the other side of it (being connected to the condenser) was a vacuum. At the end of each half cycle, this configuration was reversed, with steam pressure being applied to the other side of the piston and a vacuum created on the side opposite it. The result was a *steam* engine (atmospheric pressure was no longer involved) with double the power output of earlier models. As one might expect, the double-acting cylinder was to play a major role in the success of the steam engine. Watt's governor did not alter the way his steam engine operated, but rather the way it was controlled. An adaptation of a device already in use, it consists of two heavy balls, each anchored to a shaft by means of a hinged lever. The shaft is linked to the engine so that it rotates with it. If the engine speeds up, so does the shaft; and the centrifugal force on the balls increases so that they swing out and up. By means of a link to the throttle, the upward movement of the balls triggers a reduction in the flow of steam to the engine. And conversely if the engine slows below a certain speed.

Surprising as it may be—given the above discussion of the crank-and-connecting-rod method of producing circular motion from oscillatory motion and visa-versa, and given the wide use of such mechanisms in Britain at the time, including powering wood-turning lathes and potter's wheels—Watt did not avail himself of this method of converting the oscillatory motion of his steam engine into rotary motion. He had built a model incorporating such a mechanism, but he ended up employing a seemingly awkward system of gears to accomplish this end. However, he did not do so on the basis of technical considerations, but rather for legal and economic reasons. Boulton and Watt were prevented from thus equipping their steam engines because James Pickard had managed to secure a patent (in 1780) on the use of cranks with steam engines to convert their reciprocating motion into rotational motion. Unwilling to meet Pickard's terms and reluctant to mount a legal challenge to the patent, Watt sought an alternative technical solution to the problem, which he found in a planetary-gear arrangement. He was awarded a patent for it in 1781 and incorporated it thereafter in his steam engines when rotary power was desired. Once Pickard's patent had expired (in 1794), Watt switched to utilizing crank-and-connecting-rod mechanisms for this purpose. What appears baffling, at first glance, is that a patent was granted for something as simple and obvious as linking a steam engine to a crank by means of connecting rods to produce rotary motion when such mechanisms had long been used for such purposes in other contexts. Admittedly, just what makes an innovation sufficiently novel to merit a patent remains a matter of debate—as does the question of how long a patent or copyright should remain in force. But it seems clear enough that Pickard's patent application was an attempt to take advantage of the patent system for personal gain rather than an action by someone seeking compensation for the imagination, thought, and labor that went into creating something of value. People pursuing the former of these two courses of action have been a problem for as long as patents have been granted. Thomas Jefferson, who headed the patent office during Washington's first term as president, complained of the burden of dealing with this problem. (And things may be even worse today, as attested by the "patent trolling" phenomenon.)

The rationale for awarding patents, which are legal monopolies for specific periods of time, is to encourage people to invest the time, energy, and resources necessary for inventions to be made. The assumption is that the public will be better off in the long run if technological innovation is encouraged by shielding inventors from competition for a limited period of time, even if the short-term benefits of inventions made under such a system are reduced. But things do not always work out in the public's interest, as seen in the case of Pickard's patent, the net effect of which was to deprive the public of the benefits of a useful device for more than ten years.

And there was also a downside to Watt's patents. In 1775, he had managed somehow to obtain a twenty-five-year extension of his 1769 patent. As might be expected, he defended his patents vigorously, taking legal action against those whom he believed were infringing upon them. Unfortunately, this resulted not only in preventing the production of copies of Watt's machine but also in forestalling the production of steam engines incorporating advances devised by others. Consequently, even though Watt made some very important improvements himself, steam-engine technology changed very little in the last decade or so of the eighteenth century—especially when compared to the explosive growth of the industry, in both numbers of units and innovations, that followed the expiration of Watt's patent in 1800. With the benefit of hindsight, one is inclined to say that the public would have been better served if Watt's initial patent had expired ten years earlier. But that assumes that developments prior to 1790 would have proceeded more-or-less as they did anyway—not a safe assumption. Furthermore, to argue that the twenty-five-year extension should not have been granted fails to take into account adequately the uncertainties surrounding the project at the time and the difficulties encountered in getting the first two full-sized units to work.

4. Technology and Science

The story of the steam engine provides an excellent context in which to further discuss the role(s) of science, then and now, in the ongoing process of technological innovation. Most likely to come to mind are inventions that are more or less direct applications of scientific discoveries, e.g., radio communication. This and other examples of the handiwork of this powerful partnership, technology-and-science, are discussed below. But such is not the case with many major inventions, including the steam engine. It, like numerous others, came about as the result of efforts to find a way to deal with a practical problem: water was getting into coal mines and flooding them. As the demand for coal increased, the problem became more acute. Enterprising men began casting about for an effective way to empty the mines of water.

Those working to solve such practical problems can draw on two main types of resources: existing technologies and knowledge (as it exists at the time) of the natural order. One piece of information that proved useful to those attempting to help the coal miners had been known since ancient times: steam (produced by boiling water) exerts pressure if confined to an enclosed space. But it was not until the late seventeenth century that it became recognized that if this is the case, then a vacuum (at least a partial one) is produced when steam in an enclosed space cools sufficiently for condensation to take place. This second key feature of the natural order would have been unthinkable prior to the scientific discoveries of Torricelli, von Guericke, Boyle, et al. (see below). Its significance for the invention of the steam engine (as will also be seen below) illustrates an important role that science plays in the development of technology even when it does *not* reveal an aspect of nature that provides an immediate opportunity to advance human purposes (as did the discovery of radio waves): it provides a better—and a more powerful—understanding of the workings of nature. That is an immense advantage to those attempting to devise better ways to deal with practical problems.

As noted above, the other main resource that those seeking to achieve a practical goal must draw upon is existing technologies. The key to success may—but need not—lie in utilizing a particular one (or more) of them to exploit aspects of the natural order revealed by scientific advance(s). It may not be obvious just which strategy, involving one or more devices and/or techniques, will lead to the desired outcome. The way the two pieces of information about nature cited above might be used to remove water from mines might be thought to be obvious—but that is really hindsight. It does seem reasonable that steam pressure could be used to push water up a pipe and that the vacuum produced by cooling steam in a sealed container could be used to draw water up from a pool below. The technological resources required would appear to be quite modest: pipes, valves, an air-tight container, etc. Such seems to have been the thinking of Thomas Savery, who designed and was granted a patent in 1698 for just such a device. More accurately, in the second case he utilized the vacuum created when steam cools in a sealed container to bring about an imbalance between the gas pressure therein (near zero!) and the atmospheric pressure on the water in a pool and in the lower end of a pipe connecting the sealed container to the pool. This pressure difference forces water up the pipe. It had no moving parts except for valves, which were operated by hand. Although Savery's pump was found useful for pumping water for towns and for large buildings, it proved to be impractical for draining mines.

The path to success in that endeavor lay in utilizing a different type of mechanism to exploit what had been learned about the atmosphere, steam, and vacuums. The alternative mechanism that turned the trick was a piston moving back-and-forth inside a cylinder. Such devices had long been used as fire extinguishers, operating rather like squirt guns. In the hands of von Guericke and Boyle, they had become vacuum (or air) pumps (see below). Using hot gases under pressure to push a piston up a cylinder and allowing atmospheric pressure to push it back down when the gases cooled had been pioneered in 1680 by Christian Huygens (1629–95), who used gun powder to produce the hot gases. Ten years later Denis Papin (1647–1712) demonstrated that heating water enclosed beneath a piston in a small (three-inch diameter) cylinder produces steam that can push the piston up the cylinder and that allowing the steam to cool results in its condensation and the piston's coming back down.

Upon learning of this finding, Thomas Newcomen realized right away that since up-and-down motion is precisely the kind that is involved in the operation of a lift pump, the desired performance of such a pump could be achieved by connecting the pump rod to an oscillating piston by means of a rocking beam. Accordingly, he set about constructing a pump thus powered. Following Savery, he planned to use steam to push a piston up a cylinder and to cool it at the top of each stroke of the piston by filling a metal jacket surrounding the cylinder with cold water. The resulting condensation would allow the piston to come back down. As seen above, an accident prompted him to modify his original plan to include a device that sprays cold water into the steam-filled cylinder when the piston reaches the top of its stroke. The resulting condensation of the steam makes it possible for the piston to descend without cooling the cylinder as much as originally planned, thus reducing the amount of coal used to reheat it. It is an open question whether a pump constructed in accordance with Newcomen's original plan would have worked well enough to attract customers. This question takes on added punch if one considers that Watt later found that condensing the steam (utilized to push the piston up) in an attached container (a condenser) rather than in the cylinder itself reduces the amount of fuel used by the steam pump by a factor of one third. Put another way, this means that it took three times as much coal to operate one of Newcomen's engines than to operate one of Watt's improved versions. Furthermore, it is clear that if Newcomen's original plan (to use a cold water jacket around the cylinder itself instead of spraying a small amount of cold water into it) had been followed, the cylinder would have absorbed much more heat during each cycle, rendering the device much more inefficient—perhaps to the point that it would have been impractical.

Watt's modification of Newcomen-type engines is an excellent example of the role that knowledge of the workings of the natural order plays in technological innovation even when it does not offer an immediate answer to a problem. In a conversation with a friend, Watt recounted how he arrived at the idea of a condensing chamber separate from the cylinder of these fire engines. It is clear from his account that he had an understanding of vacuums and gases (elastic fluids such as air and steam) and that he used this knowledge in arriving at his plan to improve the performance of the machines. Because it contributed this vital information to Watt et al., it it is clear that science played an essential role in the invention of the steam engine—and thus in the launching of the Industrial Revolution, even if its part was that of a supporting cast rather than lead performer.

Nothing said above should be interpreted as suggesting some kind of dichotomy between (i) scientific advances that add to the human store of knowledge of the natural order without thereby providing (in a more or less direct way) a method of achieving a practical end and (i) those advances that clearly do provide such an opportunity. Rather, there exists a spectrum. In contrast to the supporting role of science in the iron-coal-and-steam saga discussed above, advances in science in the later decades of the nineteenth century, especially in electrodynamics and chemistry, provided unprecedented opportunities for technological innovation. An especially good example of that powerful partnership, technology-and-science, at work is afforded by the simultaneous development of organic chemistry and the chemical industry.

The story of Watt's steam engine also provides a number of examples of how those responsible for technological innovations draw upon existing devices and techniques. One that stands out in this case, the vacuum pump, was invented in 1650 by Otto von Guericke (1602–86). Like many others at the time, von Guericke was engaged in an effort to produce a vacuum, a truly empty volume. (The then-prevailing view was that it was absolutely impossible to do so.) His pump was an adaptation of a type of fire extinguisher in common use at the time. Such devices consist of a piston which can be moved back-and-forth inside a hollow cylinder by means of a handle projecting out one end of the cylinder, the other end of which is closed except for a nozzle. When the cylinder is full of water, a stream of it can be directed onto a fire by pushing the piston toward the nozzle. When empty, the extinguisher can be refilled by immersing the tip of the nozzle in a bucket of water and pulling back on the handle. This reduces the air pressure in the cylinder, and the air pressure on the water in the bucket forces it up into the cylinder. Von Guericke realized that if the nozzle of such a device were connected, in an air-tight manner, to an otherwise sealed container, it would be possible to reduce the air pressure in the container by pulling back on its handle. Furthermore, if the apparatus were fitted with an appropriate system of valves and connections, it would be possible to repeat this step and to "pump" the air out of the container, leaving a vacuum. Demonstrations that followed, showing the effects of pumping the air from various containers and hollow objects, captured the attention of the public. Audiences were shown that as air was pumped out of bell jars, candles burning under them went out. Similarly, bells enclosed under glass domes fell silent as the air was pumped out. Mice subjected to such treatment died. Probably most spectacular was von Guericke's "Magdeburg hemispheres." When the air was pumped from a hollow brass sphere formed by two hemispheres carefully fitted together, two teams of eight horses each could not pull them apart. But when a valve was opened, letting in air, the two hemispheres fell apart. Thus it was that a practical device, a fire extinguisher, was pressed into the service of science in the form of a vacuum pump and in turn incorporated into Watt's improved steam engine.

Clearly, by the mid-eighteenth century, science had become an important resource for technology, but it was not yet the powerful partner it would soon be. Indeed, not only did the steam engine not come into being primarily through the application of a scientific understanding of the physical processes involved, but it presented science with a number of tough questions: What exactly is heat, and what are the laws that govern it and its interaction with matter? Answers to these questions did not begin to assume their eventual form until the middle of the nineteenth century, but constructive steps had been taken by the end of the eighteenth century. First of all, the ancient view that fire is one of four basic substances or elements had been laid to rest. It had been established that fire or combustion is a chemical process in which oxygen combines with another element and heat and light are produced. One view of heat with a considerable following late in the eighteenth century was that heat is an invisible fluid that can neither be created nor destroyed. This "caloric" was believed to travel from hot bodies to cooler ones if they came into contact, as, for example, when a piece of hot iron is placed in a bucket of water. It was thought that caloric was released when something burns or when a solid object is sawed or drilled, which (supposedly) explains why saws and drills get hot when used for a considerable time.

This idea was put to the test by Benjamin Thompson, Count Rumford (1753–1814), when he was in charge of boring brass cannons in Bavaria in 1798. This position put at his command the economic and technological resources to conduct a wide variety of scientific experiments hardly imaginable otherwise. This example thus illustrates graphically the dependence of modern science upon its technological resources, a topic to be discussed later in the chapter along with the economic connections that go with it. Rumford had observed that the sharper the cutting instrument was, the faster it bored into the cannon stock and the more shavings were produced—and the less heat. These results were contrary to what should have happened if the caloric theory were correct. According to it, the more metal that is chewed up, the greater the quantity of heat that will be released. As a further test of the caloric theory, Rumford arranged for a very dull bit to be used in what, to all appearances, was a cannon-boring operation—except that no boring actually took place. But lots of heat was generated! He even showed that water could thus be made to boil, to the astonishment of onlookers who had never before seen water boil in the absence of fire. Of the many experiments and demonstrations that Rumford conducted, this was perhaps the most impressive. He argued that it was "extremely difficult, if not quite impossible, to form any distinct idea of any thing, capable of being ignited and communicated, in the manner that Heat was ignited and communicated in these Experiments, except it be MOTION."[1]

5. Forms of Energy

Rumford had good reason to reject the caloric theory, but his conclusion was only part of the answer. Certainly, it was not obvious how even the tiniest parts of an object could be moving with respect to each other while it remains a solid hunk of brass. In 1837 German chemist Friedrich Mohr (1806–79) offered a

1 "IV. An Inquiry Concerning the Source of the Heat which is Excited by Friction," *Phil. Trans. R. Soc.*, vol. 88, 1798, p. 99.

solution: "Heat is ... an oscillatory motion of the smallest parts of bodies."[1] An appreciation of what this claim involves and of its significance requires an understanding of the fundamental features of oscillatory motion and the implications thereof. In the most basic sense, an object can be said to be executing oscillatory motion if it just moves back and forth in a cyclical pattern. But things quickly become more complicated—and more interesting—if the discussion is focused on the characteristics of oscillating physical systems. A careful examination of these characteristics reveals that they are closely linked to principles basic to the sciences and important for technology. As a starting point, it is helpful to consider the simple case of a child moving back and forth in a swing. After passing through the lowest point of the trip at maximum speed, she begins to slow down as she ascends. Eventually she stops and begins to swing back along the same arc, picking up speed until she again achieves maximum speed at the bottom of the arc. Sweeping past it, she begins to slow down as she starts to rise once again. And so on. If further "pushes" or "pumps" are excluded and if friction is neglected in order to simplify the discussion, this example can be used to illustrate three scientific discoveries/insights that mark the beginning of a process that has had enormous technological and economic consequences. All three reside in the framework of Isaac Newton's mechanics (published in 1687), but the developments that came after them extend far beyond it.

The first of these is that moving physical objects possess "kinetic energy," which means that if they collide with another object they will strike it with force and that force will be required to slow them down or bring them to a halt. Put quantitatively, the kinetic energy of an object is equivalent to the work that would be required either (i) to accelerate it from rest to the speed with which it is moving or (ii) to bring it to a halt. In the precise terminology of physics, "work" is the magnitude of the force applied to an object multiplied by the distance over which the force acts. Such devices as sails and undershot waterwheels serve to capture the kinetic energy of wind and flowing stream so it may perform useful work. The second scientific advance referred to is the recognition of "potential energy," the ability of an object or system to do work because of its position or state. A falling rock has kinetic energy. A rock sitting at the edge of a cliff has potential energy. Should it fall off the cliff, this potential energy will be converted to kinetic energy as gravity does work on the rock. If air resistance is neglected, the amount of kinetic energy gained by the rock will be equal to the work done by the force of gravity over the distance fallen (and equals the product of these

1 "XIV. Views of the Nature of Heat," trans. P.G. Tait, *Philosophical Magazine*, series 5, vol. 2, 1976, p. 110.

two quantities). Dams, millraces, and overshot waterwheels, operating together, put potential energy to work.

The situation with the child in the swing is quite similar to that of the rock in that the potential energy associated with the swing and its occupant at their highest position is converted into kinetic energy by the time they reach the lowest point of the trip. But there the similarity ends: in the case of the swing, the kinetic energy is converted back into potential energy as the child and the swing make their next ascent. In the absence of friction, this process could go on indefinitely, with potential energy being converted into kinetic energy and that back again into potential energy. This idealized example thus illustrates the third of the scientific landmarks under discussion: the convertibility of one form of energy into another and the maintenance of the same total amount of it. This is usually referred to as the principle of conservation of energy. There is quite a variety of oscillating or vibrating physical systems, but they generally involve some kind of movement between two (or more) extreme configurations and some kind of force or forces that act to accelerate or to oppose the type of change taking place depending upon whether the system is approaching or receding from a certain intermediate position or configuration such as the low point of a swing's arc. Thus, Mohr's suggestion that heat can be understood in terms of oscillations of the smallest parts of objects requires the presence of some kind of force field that plays the same role as gravity in the swing example. These forces are in fact electromagnetic in nature, but that would not be known until the twentieth century after it was established that atoms consist of a positively charged central nucleus surrounded by negatively charged electrons.

But the experimental determination of the relationship between heat and electric current did not have to wait; it took place in the 1840s. Based on careful measurements, James Prescott Joule (1818–89) concluded that the quantity of heat generated in a given time period by an electric current flowing through a wire was proportional to the resistance of the circuit (a quantity usually expressed in ohms) multiplied by the square of the current, the rate at which the electric charge is flowing (usually expressed in amperes). This finding is referred to as "Joule's law." Joule went on to determine just how much work (expressed in terms of force times distance) was required to produce a given amount of heat. He arranged for a descending weight to turn a paddle in a container of water and measured the increase in the temperature of the water that resulted. He thus gave quantitative expression to what Rumford had observed in the cannon factory. It is called the "mechanical equivalent of heat."

Over time, more and more of these (energy) equivalences between different types of processes have come to light. Together with the principle of conservation of energy, these relationships constitute one of the most important achievements of the human quest to understand the physical universe. They have also played a key role in the emergence of the modern world, a world in which energy considerations loom so large. A crucial piece of this emerging new picture is the recognition of chemical energy, the potential for producing heat, light, motion (kinetic energy), electricity, etc. from chemical reactions. Just as a rock perched at the top of a cliff embodies the potential for producing kinetic energy, so a piece of wood or coal (carbon) together with a supply of oxygen embodies the potential for producing heat and light by way of the chemical reaction called "oxidation."

Of course, such processes as these examples have been known and exploited for a very long time. What is new is the recognition that they exhibit similarities and connections to numerous other processes in which energy is changed from one form to another but remains constant in quantity overall. In this context it is important to understand that the introduction of the term "energy" into the scientific vocabulary does not mark the discovery of a new kind of thing in the same way that the addition of the term "rattlesnake" to the language of the English marks their (first) acquaintance with that creature. Rather, the introduction of "energy" into the scientific vocabulary marks a growing awareness of one of the most basic patterns among events in the natural world: the conservation of energy, as it came to be known. The case of "energy" thus parallels those of "electricity" and "technology," as discussed in Chapter II.

Near the end of the eighteenth century, a pivotal step was taken when Alessandro Volta (1745–1827) discovered that if one makes a kind of "sandwich" by placing a small slab of spongy material saturated with a salt solution (an electrolyte) between disks of dissimilar metals and connects the two disks with a wire, a current will flow along the wire. By so doing, Volta had constructed the first **battery**, a device which converts chemical energy into electrical energy. In this case, the two disks of dissimilar metals and the salt solution in the configuration described embody the potential for causing a current to flow in a wire if it connects the two metal disks. Although it could not have been known at the time, the source of the potential energy inherent in Volta's set-up lay in the differences between the electronic structures of the atoms of the two metals. Volta also discovered that if one makes a stack of these "sandwiches," each placed with the same kind of metal disk toward the bottom of the stack (and thus each with the other kind of metal disk toward the top), and connects the disk at the bottom of the stack to the one at the top of it with a wire, a much more powerful current will flow in the wire. This device is called "**voltaic pile**" (see Figure 7-5 below). The "heads-to-tails" arrangement of the "sandwiches,"

Figure 7-5. The Voltaic Pile. Alternating layers of copper, zinc, and saltwater-soaked felt spacers provide the electrical current to two terminals. Image Credit: Luigi Chiesa. CC BY SA 3.0.

like that now utilized in ordinary dry-cell flashlights, is said to be a "series" configuration, which results in an overall voltage that is the sum of the voltages of the constituent units.

In addition to its undeniable scientific significance, Volta's work had enormous implications for technology and for economics. The great advantage of the voltaic pile over earlier sources of electricity is its ability to supply a continuous electric current. It was, of course, but the first step in the development of a major, multifaceted industry, the importance of which need not be explained to users of portable (battery-powered) electronic devices. However, much less well known is the vital role played by the voltaic pile as part of the technological repertoire of science, supplying an electric current to scientists engaged in experimental studies of one sort or another which required it, most especially those focused on the properties and effects of electric currents. Here it is important to distinguish between the voltaic pile itself being the subject of scientific investigations (for example, experiments in which Volta tried different metals, different salts, different arrangements, etc.) and the voltaic pile being used to supply an electric current for use in scientific investigations of other matters.

Remarkably, the first major discovery regarding electric currents was that they are intimately connected to magnetic fields. This finding is ironic in that one of the very first steps in understanding static electrical phenomena was to distinguish them from the effects of magnetism. In 1820 the Danish physicist Hans Christian Oersted (1777–1851) discovered that an electric current exerts a force on a compass needle, i.e., that it acts like a magnet. Almost overnight electromagnets of various configurations were being produced and experiments conducted with them. In the 1820s André-Marie Ampère (1775–1836) of France conducted detailed and systematic experiments with electric currents and formulated the results into basic laws of electromagnetism. By 1821 Michael Faraday (1791–1867) had devised an ingenious electrical circuit such that a part of it that passed through a magnetic field was free to move in a circle. And it did so when the circuit was connected to a battery. Thus, the foundation for motive power supplied by electromagnetism had been laid, but **electric motors** could not begin to assume the still-expanding central role they have until there was a much more plentiful supply of electricity than could be obtained from batteries. In 1831, following up on the idea that if electricity can produce (mechanical) motion then the reverse process should be possible, Faraday determined experimentally that an electric current can be induced in a coil of wire by subjecting it to a changing magnetic field. That condition can be achieved by thrusting a magnet through the coil of wire, by moving the coil through the magnetic field of a permanent magnet or an electromagnet, or by bringing the coil close to a circuit in which a varying electric current is flowing (such as the alternating current now in common use). This process is called "**electromagnetic induction.**" Other scientists and inventors quickly devised **electric generators** capable of producing usable, non-transitory electric currents.

6. Dividends of the Powerful Partnership: Illumination, Communication, and Power Production

The voltaic pile and the scientific advances that followed enabled the invention and widespread use of an entire array of devices and processes. One of the most important of these is electric lighting. Few technological innovations have so fundamentally transformed the daily rhythms of peoples' lives as has the ready availability of **artificial lighting.** Prior to this, the nighttime activities of common people (especially) were so severely restricted that their day-to-day lives are now unimaginable. In 1800 people still relied on candles, lamps fueled by animal fats, torches, and open fires for illumination. In North America, whaling (to obtain oil for lamps) was a major industry. This situation changed dramatically during the next century as one innovation after another united in pushing back the darkness.

The first of these was gaslight. Although many had experimented with the flammable gases known to be released from coal when it is heated to make coke, William Murdoch (1754–1839) took things a step further and turned it into a major source of illumination. An inventive and highly competent engineer (though with little formal education), Murdoch was overseeing the erection and operation of Boulton and Watt steam engines in Cornwall when he devised a way of lighting his home with coke-oven gas. With the technical resources available to him when he returned to the Soho Works in 1798 as manager, Murdoch pressed forward with the development of gas lighting. This endeavor culminated in the illumination of the exterior of the factory in 1802 as part of a national celebration, an achievement that attracted wide attention. Almost overnight city streets were lit by gaslight, as were factories where operations were no longer confined to daylight hours. Later in the century, it was found that a similar gaseous fuel ("natural gas") could be obtained from wells. Unfortunately, especially in the early years, people could not enjoy the benefits of gaslight unless they lived near gas wells or a gas plant.

This limitation on artificial lighting was overcome thanks to the discovery of kerosene in 1852 by Abraham Gesner (1797–1864), a Canadian physician and geologist. Commonly called "lamp oil" (because of its use), kerosene was first distilled from coal (hence the name "coal oil"), but it soon came primarily from petroleum refining operations. As kerosene became available for use in lamps, the American whaling industry went into decline. Kerosene continued to be used to provide light in rural areas and isolated locations well into the twentieth century. But even before kerosene was discovered, the epitaph for its reign was already being drafted. In 1831, as discussed above, Michael Faraday had established that an electric current can be induced in a coil of wire by subjecting it to a changing magnetic field. Faraday's discovery provided the foundation for a major enterprise, the generation of electric current, which developed rapidly. Very early generators were hand cranked, but soon larger ones were being powered by small steam engines. By the middle of the century, electricity was being produced in greater and greater quantities and used for a growing variety of purposes. Initially illumination was provided by arc lights (light produced by current flowing across a gap between two electrodes), which were of limited usefulness as they were suitable only for street lights, theaters, light houses, and the like. Only after the invention of the incandescent light bulb by Thomas Edison (1847–1931) in 1879 did electric lighting become available for homes, small shops, etc.

Much earlier—even before electricity became generally available from generators—important applications were already being developed, especially in communications. The earliest device for sending messages by electrical signals, the telegraph, appeared in 1837 after being developed by English entrepreneur and inventor William Cooke (1806–79) and English scientist Charles Wheatstone (1802–75)—yet another instance of science and technology advancing hand-in-hand. Cooke and Wheatstone's telegraph involved a multi-wire apparatus and a letterboard that made it possible to transmit messages that directly encoded letters selected by an operator, with no need for the operator to learn the code. However, a single-wire telegraph was developed by Samuel Morse (1791–1872) in the United States that required the operator to learn a special code, and six years later the first message was transmitted by Morse code. The single-wire design was more economically efficient, requiring fewer wires for transmission and maintenance, and while Cooke and Wheatstone's telegraph was still used in England, economic pressure eventually caused them to develop a single-wire system as well.

The transmission of the human voice by means of electrical currents, that is, the telephone, invented by Alexander Graham Bell (1847–1922), followed in 1876. Meanwhile James Clerk Maxwell (1831–79) pulled together the then-known laws of electromagnetism and made an important modification to one of them. The mathematical expression of the resulting (four) laws is known as "Maxwell's Equations." At the time, a main focus of physicists' attention was the question of the nature of light. In 1862 Maxwell advanced the theory that light is an electromagnetic phenomenon, governed by these four laws. Although he did not then devote himself to developing and checking up on the implications of this theory, others did. In 1883 George FitzGerald (1851–1901) noted that it follows from Maxwell's theory that an electric current rapidly surging back and forth in a circuit would be expected to emit electromagnetic radiation, and he suggested circumstances under which this would take place. Following up on FitzGerald's suggestions, Heinrich Hertz (1857–94) confirmed the existence of these "radio waves" three years later. By 1894 Guglielmo Marconi (1874–1937) had succeeded in transmitting and receiving signals by means of them. The age of wireless transmission of information had arrived.

Motive power is the third major area in which applications of electromagnetism were developed once electricity was generally available. One of the more impressive immediate effects of the availability of electric motors was to banish from factory floors the elaborate—and dangerous—systems of belts, pulleys, and drive-shafts once used to power machines. With electric motors to drive the individual machines, there was no need for these cumbersome networks. The great variety of electrical appliances, tools, and instruments so pervasive in modern life came later when small, fractional horsepower electric motors were available. Many uses were (and continue to be) found for electromagnets, especially solenoids with a ferrous metal core free to slide back and forth inside a cylindrical coil of wire. These devices are used to operate switches, valves, latches, etc.

Unlike the advances in ironworking traced above, the invention of the steam engine, and the discovery of important derivatives to be had from coal, in which science played no role at all or only a secondary one, the invention and development of electromagnetic illumination, communication, and power production were directly dependent upon scientific discoveries, as outlined above. And as noted in Chapter II, the case of the chemical industry is similar: it blossomed only after foundations were laid for it by science. Given the dividends paid by the powerful partnership of science and technology, not only during the nineteenth century but ever since, it can hardly be surprising that so many people once accepted the view that technology, properly understood and practiced, is "applied science" and that most people do not distinguish between technology and science in a clear and consistent way. Some, no doubt, will regard any attempt to differentiate between them as an exercise in "hairsplitting."

But the distinction between science and technology is important when making decisions regarding investments or funding for proposed projects. In assessing the viability of a scientific research program (which may or may not involve the development and implementation of novel technological resources), it is important to keep in mind that scientific investigations, even when successful from a scientific point of view, do not necessarily yield anything of immediate practical value as regards food, shelter, medicine, energy, etc. Even so, the results thus obtained, as well as any new devices or processes developed, become part of the resources available to subsequent investigators and problem solvers. But there is no guarantee that any particular item contributed by science to this treasure trove will ever serve a practical purpose, nor is there any general and reliable method of predicting such usefulness. Furthermore, one must take into account the intrinsic value of understanding ourselves and our world. Consequently, addressing questions regarding the allocation of funds to support scientific research is a tricky affair. On the other hand, most technologies, intended as they are for practical purposes, must be assessed in terms of their costs, their benefits, and the likelihood of these being realized. And this is true in the case of an entrepreneur contemplating launching a new product as well as in the case of an organization deciding whether to purchase a new type of equipment. Thus, the final part of this chapter is concerned with the impact of economic concerns on the relationship between technology and science, especially in the modern world.

B. Economics and the Technology of Science

As a result of its dependence upon its technological resources, modern science is tied to economic factors in ways that have not always been appreciated. Whether or not a line of research is pursued often depends upon the cost of the equipment needed and availability of funds to pay for it. As the level of scientific activity has increased and as the instruments employed have become more complex and specialized, researchers have come to depend more and more upon instruments and machines supplied to them by engineering firms and manufacturers. Consequently, experimental scientists who work with equipment of this kind and who conceive of ways that it can be improved will need to convince the makers of the relevant kinds of equipment of the merits thereof and to persuade them to invest in the proposed innovations. Such suggestions may, of course, come from industrial workers with science backgrounds who use certain kinds of instruments. But the managers of the businesses that make such products can be expected to make their decisions on the basis of the likelihood that such ventures will be profitable. Thus, should it be the case that a proposed innovation holds the potential for opening up an entirely new line of research, it is possible that a business decision, made for economic reasons, could delay or even prevent research of a certain kind from being undertaken. But even repeated rejections by relevant firms need not be the end of the story. Scientists who are convinced of the merits of their idea sometimes strike out on their own and succeed at establishing a viable production operation. That is exactly what happened in the case of the **quadrupole gas chromatograph/mass spectrometer**—with seismic results in both chemical science and in industry.

1. The Quadrupole Gas Chromatograph/Mass Spectrometer

After receiving his Ph.D. in electrical engineering in 1957, Robert E. Finnigan (1927–2022) was employed as a senior scientist at the University of California's Lawrence Livermore National Laboratory, where he led an effort to develop computer-operated instrumentation and control systems for nuclear reactors. In 1962, he was recruited, along with co-worker Mike Uthe, by the Stanford Research Institute to launch a controls program there. Serendipitously, at this very time another team at that institute was at work creating the first quadrupole mass spectrometer. With the advantage of their experience with computer applications, Finnigan and Uthe concluded right away that the device had tremendous potential, but did not succeed in attracting commercial interests. Two years later, Finnigan and Uthe joined Electronic Associates, Inc., a leading supplier of analog computers, as founders of its Instruments Division. The division began operating as a commercial manufacturer in 1964. By 1966 it had sold over 500 quadrupole residual gas-analyzer instruments, many of them for use in microelectronics and materials science research.

By the mid-1960s mini-computers had become available and were being put to work in research laboratories. These machines were about the size of a refrigerator. (That may seem "huge" by current standards, but earlier computer installations took up very large rooms or even entire floors.) They cost several tens of thousands of dollars. It is not surprising that in such a context Finnigan's thoughts focused more and more on the possibility of combining a gas chromatograph and a quadrupole mass spectrometer into a single unit under the control of a mini-computer. Finnigan knew that the quadrupole mass spectrometer was especially suited for control by a mini-computer because of the way it operated. And he believed that such a combined unit would not only facilitate and expedite data collection but would also open the door to a whole new range of possibilities as regards data analysis—in short, speed and power. But Finnigan was not able to convince the management of Electronic Associates, Inc. to underwrite the proposed project. Undaunted, he set about forming the Finnigan Instrument Corporation to produce the machines. In 1968 the company delivered two prototype quadrupole gc/ms (gas chromatograph/mass spectrometer) instruments, one to a biochemistry laboratory at Stanford University and the other to the chemistry department of Purdue University. But the following year the company filled only one order. The high price, $100,000 (over $600,000 today) was surely partly responsible. And it would take time for the capabilities of the new device to become known. Unless potential customers could be convinced that such a large investment would pay dividends, the curtain would come down quickly on Finnigan's venture.

Again serendipitously, developments unrelated to his current focus presented Finnigan and co-workers with an opportunity. In 1970 the Environmental Protection Agency was created and charged with establishing standards and procedures to be used by government, industry, and courts in monitoring pollutants and in enforcing the laws and regulations pertaining to them. Finnigan recognized right away that the presence of the organic compounds that most concerned the E.P.A. could be identified and quantified reliably and quickly by the quadrupole gc/ms instruments made by his company. He and co-workers launched an intensive campaign to persuade E.P.A. officials of the merits of their product. In 1971 the E.P.A. placed an order for 20 of the devices. They quickly demonstrated their worth with respect to both speed and accuracy. Analyses that previously might have taken months could be completed in a matter of hours, and findings obtained by means of them came to be accepted as reliable evidence in law courts. And, as seen in the case study of the instrumental revolution in Chapter II, the devices were put to work for government and industry in rapidly growing numbers. Thus, this complex instrument developed to serve scientific purposes first gained acceptance for its practical usefulness.

This account of the development of the quadrupole gas chromatograph/mass spectrometer illustrates forcefully several points made above and earlier in Chapter II regarding science and technology. Firstly, the porous and diffuse nature of the border between the two is very much in evidence in the odyssey of Robert Finnigan. His undergraduate and graduate degrees were in electrical engineering, and he was first employed to design and construct instruments and computerized control systems for nuclear reactors. Because of his experience with computerized control systems, Finnigan was able to recognize opportunities to greatly enhance the speed and power of particular kinds of scientific instruments by placing their operation under the control of computers. Even though established manufacturers of scientific and technical equipment declined to underwrite one and then another such venture, Finnigan had sufficient understanding of the sciences, especially chemistry and biochemistry, to be confident of what the devices could do for them. Accordingly, in each case, he launched a business to produce the instruments, with his first deliveries going to scientific research groups.

Secondly, this episode demonstrates how the dependence of science upon its technology can determine how it evolves. The breakthrough for Finnigan's crowning achievement, the quadrupole gas chromatograph/mass spectrometer, and for his business ventures came from its practical application. Only later did it join other technological innovations in revolutionizing the practice of chemistry. Thirdly, the example of the practical application of the quadrupole gc/ms device illustrates how efforts to advance science by enhancing its technological resources can lead to new technologies coming into widespread practical use. And lastly, the challenges encountered by Finnigan in his efforts to get the quadrupole mass spectrometer into production illustrate the relevance of economic considerations to the course that science takes. Eventually it was recognized that the capabilities of these machines justified their cost. But that does not mean that there is some kind of guarantee that costs of the equipment needed to pursue a certain line of research will not stand in the way of its being done. There could hardly be a more dramatic example of costs bringing a scientific research endeavor to a halt than the cancellation in 1993 of the Superconducting Supercollider. At that point two billion dollars had already been spent on this giant particle accelerator, the total cost of which was estimated to have been 8.5 billion dollars.

2. Laser Cooling and Trapping and the Bose-Einstein Condensate

Economic considerations related to the technology of science are not always a hurdle in the path of scientific inquiry. Quite the contrary. Interests in the practical applications of a device or a process being developed sometimes provide the economic support needed to move the project along. A striking example is **laser cooling with tunable diode lasers.** Ironically, this episode begins with a proposed research project that would have required a million dollars just for the required lasers, a sum that proved to be unattainable. Faced with this challenge, experimental physicist Carl Wieman (b. 1951) set about devising a way to render the relatively cheap diode lasers (as used in CD players) tunable, i.e., to make it possible to vary their frequencies (roughly, the color of their light). Once he had accomplished this, Wieman began to consider the avenues that might be explored now that the cost of doing so was not so prohibitive. What followed is one of the most powerful demonstrations imaginable of what can happen when scientists are encouraged to try things they would not have tried had not the costs of doing so been so drastically reduced.

Wieman and co-workers decided to try their hand at using lasers to slow atoms down. This can be accomplished by taking advantage of the quantum mechanical nature of the way light interacts with atoms. It is thus possible to arrange things so that an atom receives a head-on "whack," slowing it down, if it is traveling toward a laser when an interaction takes place. On the other hand, the arrangement is such that a much weaker or no interaction will take place if an atom is moving away from the laser. Since slowing atoms with lasers had been achieved already using the older, expensive lasers, the real step forward taken by Wieman was showing that it could be done with lasers costing only a tiny fraction (about one thousandth!) of the earlier price. Wieman and a co-worker then turned their attention to **laser trapping**, not merely slowing atoms with lasers but confining them to one spot by means of lasers on all sides, again a feat that had been accomplished earlier with the older, expensive lasers.

Although they were unsuccessful at obtaining a government grant to support their envisioned investigations of atomic phenomena at the extremely low temperatures they hoped to attain with their new technology, Weiman and co-workers did receive funding from the Office of Naval Research. Very accurate clocks being essential to navigation and extremely low temperatures being relevant to the accuracy of atomic clocks, the Navy provided financial support for Wieman and co-workers to push the new laser-cooling technology to reach lower and lower temperatures. Soon they were achieving temperatures that were orders of magnitude lower than any previously recorded. (Perhaps it should be noted that the reference here is to decimal fractions of a degree above absolute zero.) As this endeavor progressed, these physicists began to notice unanticipated atomic-level phenomena, atoms interacting in ways previously unknown, and even atoms being trapped under conditions not expected to have permitted it. All of which motivated them to press on. This they did, eventually bringing about a weird new state of matter called a Bose-Einstein condensate, which was predicted by quantum mechanics. Bose-Einstein condensates are created by cooling low-density gases of bosons to nearly absolute zero, which allows certain kinds of (typically microscopic) quantum phenomena to have macroscopically observable effects. For this achievement, they were awarded the 2001 Nobel Prize in Physics.

This episode aptly illustrates the economic aspects of scientific research, especially as regards the instruments devised by scientists in pursuit of their quests. Of course, scientific research requires economic support, but the notable—but by no means new or unique—feature of this example is that the crucial economic support came from those whose interest was in the technology being developed rather than the scientific goals of the project. A second important feature of this example is that the availability of these relatively inexpensive tunable diode lasers made feasible an entire array of research programs that simply would not have been undertaken in their absence because of the costs involved. As a result of their invention, not only were Wieman and co-workers able to produce a Bose-Einstein condensate (even though they had not originally set out to do so), but atomic physics, in the doldrums for decades, quickly became an extremely active research area. There could hardly be a clearer example of the influence of economic considerations upon the direction taken by scientific research or of the critical role of technology in linking the two.

Sources / Further Reading

Ashton, Thomas S. *The Industrial Revolution: 1760–1830*. Oxford University Press, 1964.

——. *Iron and Steel in the Industrial Revolution*. Manchester University Press, 1963.

Beauchamp, Ken. *History of Telegraphy*. London: The Institution of Engineering and Technology, 2001.

Brock, David C. "Tools and Technology: A Measure of Success." Science Heritage Institute. 6 June 2011. sciencehistory.org /distillations/a-measure-of-success

Bronowski, Jacob. *The Ascent of Man*. Little, Brown, 1973.

Burke, James. *Connections*. Boston: Little, Brown, 1978.

Canby, Edward Tatnall. *A History of Electricity*. Hawthorn Books, 1963.

Clark, Gregory. *A Farewell to Alms: A Brief Economics History of the World*. Princeton University Press, 2007.

Clark, Ronald W. *Works of Man*. Viking, 1985.

Craughwell, Thomas J. *The Book of Invention*. Black Dog & Leventhal, 2008.

Davis, Bob, and David Wessel. "At Dawn of Electricity, Feuds and Hype." *Wall Street Journal*, vol. 330, no. 66, 6 April 1998, p. A17.

Dunsheath, Percy. *A History of Electrical Engineering*. Pitman, 1962.

Dyson, James. *A History of Great Inventions*, edited by Robert Uhlig. Carroll & Graf, 2001.

Evans, Harold, et al. *They Made America: From the Steam Engine to the Search Engine, Two Centuries of Innovators*. Little, Brown, 2004.

Finkelstein, Joseph, editor. *Windows on a New World: The Third Industrial Revolution*. Greenwood Press, 1989. Contributions in Economic and Economic History No. 88.

Franklin, Alan. *The Neglect of Experiment*. Cambridge University Press, 1989.

Galison, Peter. *Image and Logic: A Material Culture of Microphysics*. University of Chicago Press, 1997.

Gies, Frances, and Joseph Gies. *Cathedral, Forge, and Waterwheel: Technology and Invention in the Middle Ages*. HarperCollins Publishers, 1994.

Gimpel, Jean. *The Medieval Machine: The Industrial Revolution of the Middle Ages*. Holt, Rinehart and Winston, 1976.

Harris, Daniel C. *Quantitative Chemical Analysis*, 6th ed. W.H. Freeman, 2003.
Hodges, Henry. *Technology in the Ancient World*. Alfred A. Knopf, 1970.
Klemm, Friedrich. *A History of Western Technology*. Translated by Dorothea Waley Singer. MIT Press, 1964.
Locke, Robert R. *The End of the Practical Man: Entrepreneurship and Higher Education in Germany, France, and Great Britain, 1880–1940*. Greenwich, Connecticut: JAI Press, 1984.
Magie, William Francis, editor. *A Source Book in Physics*. Harvard University Press, 1963.
March, Raymond E., and John F.J. Todd. *Quadrupole Ion Trap Mass Spectrometry*, 2nd ed. John Wiley, 2005.
Mason, Stephen F. *A History of the Sciences*, revised ed. Collier Books, 1962.
McFadden, William. *Techniques of Combined Gas Chromatography/Mass Spectrometry: Applications to Organic Analysis*. John Wiley, 1973.
Morris, Peter J.T. *From Classical to Modern Chemistry: The Instrumental Revolution*. Cambridge: The Royal Society of Chemistry, 2001.
Patton, Phil. *Made In U.S.A.: The Secret Histories of the Things That Made America Great*. Grove Press, 1992.
Petroski, Henry. *Remaking the World: Adventures in Engineering*. Alfred A. Knopf, 1997.
Pitt, Joseph C. *Galileo, Human Knowledge, and the Book of Nature: Method Replaces Metaphysics*. Dordrecht: Kluwer Academic Publishers, 1992.
Ritti, Tulia, Klaus Grewe, and Paul Kessener. "A Relief of a Watered-powered Stone Saw Mill on a Sarcophagus at Hierapolis and Its Implications." *Journal of Archaeology*, vol. 20, 2007, pp. 139–63.
Robinson, Eric, and Douglas McKie. *Partners in Science: Letters of James Watt and Joseph Black*. Harvard University Press, 1970.
Ronan, Colin A. *Science: Its History and Development Among the World's Cultures*. Facts on File, 1982.
Rosenberg, Nathan, and L.E. Birdzell, Jr. *How the West Grew Rich: The Economic Transformation of the Industrial World*. Basic Books, 1986.
Rothbart, Daniel. *Philosophical Instruments: Minds and Tools at Work*. University of Illinois Press, 2007.
Saugus Iron Works 1647: The First Successful Iron Works in America—A National Historic Mechanical Engineering Landmark. Saugus, Massachusetts: National Park Service and The American Society of Mechanical Engineers. June 15, 1975. https://www.asme.org/wwwasmeorg/media/resourcefiles/aboutasme/who%20we%20are/engineering%20history/landmarks/7-saugus-ironworks-1647.pdf
Seymour, John. *The Forgotten Crafts: A Practical Guide to Traditional Skills*. Alfred A. Knopf, 1984.
Silver, Brian L. *The Ascent of Science*. Oxford University Press, 1998.

The Smithsonian Book of Invention. Smithsonian Exposition Books, 1978.

Stark, Rodney. *How the West Won: The Neglected Story of the Triumph of Modernity.* Intercollegiate Studies Institute, 2014.

Turner, Gerlad L'e. *Scientific Instruments 1500–1900: An Introduction.* University of California Press, 1998.

Wertime, Theodore A., and James D. Muhly, editors. *The Coming of the Age of Iron.* Yale University Press, 1980.

White, K.D. *Greek and Roman Technology.* Cornell University Press, 1984.

Wieman, Carl E. "Carl E. Wieman—Biographical." Nobelprize.org. Nobel Prize Foundation, 2001. https://www.nobelprize.org/prizes/physics/2001/wieman/biographical/.

Wieman, Carl E., and Gwenn Flowers. "Inexpensive Laser Cooling and Trapping Experiments for Undergraduate Laboratories." *American Journal of Physics*, vol. 63, no. 4, April 1995, pp. 317–30.

Wieman, Carl E., et al. "A Narrow-band Tunable Diode Laser System with Grating Feedback, and a Saturated Absorption Spectrometer for Cs and Rb." *American Journal of Physics*, vol. 60, no. 12, Dec. 1992, pp. 1098–1111.

Wieman, Carl E., and Leo Hollberg. "Using Diode Lasers for Atomic Physics." *Reviews of Scientific Instruments*, vol. 62. no. 1, January 1991, pp. 1–20.

Williams, Trevor. *A History of Invention: From Stone Axes to Silicon Chips*, revised ed. Updated and revised by William E. Schaaf, Jr. Checkmark Books, 2000.

Wilson, Andrew. "Machines, Power and the Ancient Economy." *The Journal of Roman Studies*, vol. 92, 2002, pp. 1–32.

Chapter VIII:
The Interplay between Social and Political Factors and Technology

GIVEN ITS INTIMATE TIES to economic matters, technology could hardly fail to be closely related to social conditions and developments. As discussed in Chapters I and II, it was the social disruptions and deprivations associated with the Industrial Revolution that occasioned the social criticism in which one finds the dawning recognition of the importance of technology—indeed, the recognition of the significance of such a category. Technology may not have assumed a place in the public consciousness until the twentieth century, but early in the nineteenth century the real-life economic significance of technological innovation was brutally obvious to the skilled textile workers of England. As new power-driven machines, tended by low-paid operators with minimal training, were put in operation turning out cloth more cheaply and in greater quantities, traditional craftspeople realized that the value of their special skills and their very livelihoods were threatened. Up to that point, they had enjoyed a recognized position in society and the assurance of a modest but adequate income. Secretly organized (trade unions were prohibited), the workers at first attempted (with partial success) to prevent the installation and operation of the new machines through negotiation, disguised strikes, and threats. When efforts to gain relief legally were rebuffed, the Luddites, as they came to be called, turned to smashing the new textile machinery, escalating their threats against those who put it in operation, and even attempting to storm heavily defended factories. As violence escalated, it seemed for a time that England was teetering on the brink of an open insurrection. But the Luddites lacked any kind of central organization or coherent agenda for reform. Thus, following repeated hammerings by private guards, the police, and the army, the restive workers dispersed and things quieted down. Even so, there was overwhelming support for the Luddites among the poorer classes and a growing general recognition of the need for reform. In the decades following the end of the Napoleonic Wars, a number of reform measures were passed, including (in 1824) the repeal of laws prohibiting the formation of trade unions. The new production methods had brought an end to a way of life, and a new era was beginning.

A. Technology and Society: Perspectives and Approaches

Few people today would choose to meet their basic needs for clothing, bed linen, towels, etc. by selecting only items made from hand-woven fabric, given the price these would command currently. But the fate of the displaced weavers, croppers, and other cloth-workers is troubling—especially so since this same pattern of events has shown itself repeatedly from the onset of the Industrial Revolution right up to the present. Anyone seriously concerned with matters related to the waves of innovation currently sweeping through the information and communication sector (and other areas affected) must examine at least some of the more notable such developments in the past with an eye to taking advantage of whatever insights that may be gained thereby.

To anyone born after 1970 or thereabouts, the presence of continuous and rapid technological innovation—and the accompanying obsoletion of devices, techniques, and skills—is simply taken for granted. But things were not always so. No one who has studied the topic can fail to be struck by the difference between conditions prior to the Industrial Revolution and those that prevailed after it got underway. Perhaps most notable is the speed at which changes induced by technological innovation began to take place. During the earlier period, such changes typically came about at such a glacial pace that little difference would be noticed during one person's lifetime. In contrast, the years that followed witnessed the disappearance, almost overnight, of trades that had sustained workers for generations. As might be expected, those concerned with the social disruptions accompanying the new modes of production as well as those inclined to celebrate their power and efficiency have sought to identify the decisive factor(s) involved in this great watershed in human affairs.

It might seem, as some have argued, that the real culprit in cases like that sketched above is not technology but the economic and political system (e.g., industrial capitalism) in which it is implemented. In generalized form, such a view is called "instrumentalism." According to it, technologies are (in simplest terms) just tools, ways of getting things done, and any good, ill, or injustice that results from their use derives from how they are used or the ends they are deployed to promote. Thus, on this view, whether the benefits of an invention are widely distributed or bestowed only upon a few depends upon the economic and political system involved. A simple but graphic example is provided by the conditions surrounding the boom in the construction of gristmills in England during the late Middle Ages. Because of its topography and climate, England is an ideal location for overshot waterwheels, which are much more powerful and efficient than other types but are also much more costly to build (as discussed in the preceding chapter). In the feudal system of the time, those able to finance such an undertaking almost certainly would have been members of the nobility or members of a religious order with access to its wealth. As one might expect, the proprietors of these mills pursued every method at their disposal to realize a profit on their investment. One widely employed strategy was to require serfs to bring their grain to their lord's mill for grinding—even though the serfs would have preferred to avoid this expense by grinding their grain on their own hand-operated querns. Doing so would have also saved them the time and effort required to transport their grain to the mill and the flour back home, as well as the time spent waiting in line at the mill. To make matters worse, the mill owners charged the peasants bound to their estates at twice the rate they charged those who could not be forced to patronize their mills. Those customers were even allowed to "jump the queue" at mills for a fee.

Evidently the injustices accompanying the proliferation of grain mills in England during the late Middle Ages derive from the feudal system of the time, not from the technology itself. After all, the water-powered gristmill represents the final liberation of human beings from the laborious, mind-numbing chore of manually grinding grain—a task as old as agriculture—usually performed by women. In contrast, while the mechanization of the spinning and weaving process did provide people with cheaper cloth and the factory owners with profits, it wiped out the economic value of the special skills of the traditional textile workers and left them with few prospects for the future. Evidently, in this case, some of the responsibility for the injuries suffered by the traditional textile workers must be assigned to the new technology. Consequently, it cannot be the case that the effects of the utilization of a technology are good, bad, unjust, etc. solely because of societal factors. Indeed, at the opposite end of the spectrum from the liberating grain mill are instruments of torture for which, arguably, there is no defensible use under any conditions. (There is a disturbing display of these devices in the museum of Hohensalzburg Castle in Austria.)

Among those who oppose the thesis that technology is neutral with regard to benefits, harms, injustices, etc. are social scientists and historians who subscribe to what is called "social contextualism" or "social determinism." These thinkers point out that a technology can hardly be said to be "neutral" if it reinforces markedly the economic and political power of those who control it and introduces changes that are unwanted by a majority of the population. Social contextualists argue that the make-up of technologies derives not only from technical considerations of functionality but also from a variety of political, social, and cultural factors, including value systems. For example, the path that technological development takes (and consequently the make-up of the technological repertoire of a society or culture) is determined (at least in part) by decisions (often self-serving) made by people in positions of power. Thus, supposedly, technologies are not self-contained tools or systems that can exist and function separately from and independently of the social and cultural matrix of which they are a part.

Certainly the social contextualists have a point. All technologies do exist in some social and cultural context, as is implied by the definition worked out in Chapter II with its emphasis on use. And it would be difficult to deny that the factors which the social contextualists emphasize determine (or help to determine) the time and place that particular inventions are made or successfully implemented (or where such attempts fail). Furthermore, these factors can account (to various degrees) for many things: the uses to which technologies are put or adapted, variations in the forms that they assume, the patterns that technological development takes, the rate at which technological innovation takes place, and even whether it takes place at all. But it would be a mistake to think that technologies cannot be appropriated by people living in societies that are radically different from those in which the technologies are already established. Indeed, it is quite common for such transfers of devices and techniques to take place. A good example is the introduction of firearms into feudal Japan, discussed fully in the case study below. Contrary to what some extreme forms of social contextualism might seem to suggest, technologies can exist and function in the absence of many, if not most, of the cultural and societal elements of their place of origin.

Those who emphasize the social conditioning or shaping of technology do, however, seem to be on target with their criticisms of "technological determinism," according to which technology, rather than being guided by human goals and decisions, is an autonomous entity that follows its own course of development determined by its own internal logic. This view has a certain initial plausibility. As seen in earlier chapters, technologies have repeatedly ushered in unanticipated and unintended changes in the very nature of human life. Although technological determinism has many articulate proponents, its critics have called attention to a number of cases in which powerful technologies have failed to win acceptance, have been proscribed or tightly restricted, or simply have not been utilized. The reasons for these differing outcomes include religious commitments, value systems, cultural traditions, strategies and policies, etc. For instance, some Old Order Mennonites and Amish refuse to use not only electricity and automobiles but also modern sanitation methods. Other examples often cited include belligerents refraining from using poison gas in the Second World War, the virtual elimination of firearms from Japan for a 250-year period beginning in the seventeenth century (even though they had demonstrated their superiority in numerous military campaigns), and the banning of the insecticide DDT in the United States and most other industrialized countries (while its value in the battle against malaria in tropical lowlands is still being debated).

In addition to determining whether or not particular technologies are put to use, social forces can heavily influence the forms that technologies take and can, in general, encourage or discourage technological innovation. This means, of course, that how a society responds to a technology will affect its impact on that society. Human societies and their relations to technologies are very complex, however, so that it is often very difficult, if not impossible, to determine precisely which factors are involved in a given case and how they conspire to produce a specific effect. Even when it may seem clear that a particular social or political factor is responsible for a certain situation or result, further investigation may indicate that others may be involved. Because of such complexities and the bewildering variety of devices and techniques that human beings have evolved, a systematic and encompassing treatment of questions like those just mentioned is not in the offing.

Some insight into of the mutual dependence of technology and human societies can be gained, however, by focusing upon individual cases. For example, by studying the developments that follow the introduction of the same technology into very different societies under different circumstances, one may be able to discover unsuspected correlations or dependences or find that expected ones are absent. An excellent example for such purposes is afforded by **gunpowder.**

Invented in China, it was introduced into Europe and Japan with very different results. It first emerged in a series of steps extending over several centuries, and—contrary to a widespread misconception in the West—the Chinese were quick to develop military applications of the "fire drug," as they called it. But revolutionary changes—in warfare or in society—did *not* ensue in China. In contrast, gunpowder appeared quite suddenly in Europe; and long before a century had elapsed, cannons were being fired at the walls of European towns and castles, rendering such defensive measures obsolete almost overnight. Not only did gunpowder trigger fundamental changes in Europe, but it played a key role in the extension of European hegemony over much of the planet.

In Japan there occurred a very different turn of events: a reversion to edged weapons followed, in less than a century, the introduction of firearms into the country by Europeans in the 1540s. Within a few years of the arrival of the Europeans, skilled Japanese craftspeople were turning out firearms by the thousands. However, the production and use of guns and gunpowder began to decline early in the 1600s following the Tokugawa shogunate's taking control of the manufacturing of firearms in its bid to consolidate power. Over time, firearms gradually faded from the scene, not to reappear in Japan for over two hundred years.

These three very different scenarios provide an opportunity to reflect upon the differing effects that a technology can have on a society, the factors that can influence what these effects are, as well as those factors that affect the course of development of a technology. Accordingly, a case study focusing on these three episodes is included in this chapter. Because industrialization has had such an impact on the lives of women and on family life, a section devoted to that topic follows the case study. In the final section of the chapter, themes introduced earlier are explored further.

B. Case Study: Gunpowder, Pyrotechnics, and Samurai Swords

1. The Chinese and Gunpowder

Invented by the Chinese (probably in the ninth century), gunpowder seems to have emerged from experiments carried out by Daoist alchemists in their pursuit of an elixir of immortality. Indeed, the first clear evidence of the preparation of a mixture containing the essential ingredients of gunpowder is an ancient Daoist text warning of its dangers: smoke, flames, burned hands and face, even buildings set aflame. Thus, one might say that the first 'identity' of gunpowder was a dangerous mixture of alchemical materials that was to be avoided. Early forms of what the Chinese called the "fire drug" were highly incendiary but not explosive. Because of the ceremonial and entertainment value of impressive bursts of flame and sudden loud noises, the fire drug soon found pyrotechnical applications. As a result, gunpowder assumed a second identity: the fire drug that makes fireworks possible, the fuel for a national passion in China that has lasted for a thousand years. The explanation that seems to have been favored by the Chinese for the awesome properties of the fire drug was that it brought into close proximity the polar opposites yin and yang. While this account calls to mind the modern conception of what happens when matter and anti-matter are brought together, yin and yang are not material elements. They are perhaps best thought of as abstract categories underlying every possible duality, complementarity, or cyclic change. Even so, in some accounts they seem to function as forces, powers, or agents.

The popularity of pyrotechnics, as well as other uses (see below), seems to have encouraged efforts to produce more and more powerful mixtures, culminating in an explosive substance: gunpowder, as it came to be known. And, contrary to widespread belief in the West, the Chinese were quick to exploit the military potential of the fire drug, beginning with its incendiary properties, which rendered it eminently suitable for use in warfare. The Chinese had long used fire arrows as weapons, and an arrow carrying a packet of gunpowder with a lit fuse is much more effective at setting a fire that is difficult to extinguish than is an arrow armed with burning pitch or a similar combustible. By the eleventh century, the Chinese were making gunpowder fire arrows by the tens of thousands.

Even relatively low-grade gunpowder can be used to produce an explosion if it is ignited in an enclosed space. Combustion can take place under such circumstances because gunpowder carries its own oxygen supply with it, locked up in solid form in the saltpeter (potassium nitrate or KNO_3). The immediate availability of oxygen in gunpowder (which must be well mixed) accelerates combustion and the production of gases, which, in a confined space, produces a sharp increase in pressure. Under the right conditions, the pressure can be sufficient to burst the container. The result is a loud report and a cloud of thick, white smoke. Early experiments with bamboo or paper and string "firecrackers" paved the way, as the quality of gunpowder improved, for cast iron bombs that burst with deadly force, spraying adversaries with shrapnel. Early thirteenth-century records describe the devastating effects of these iron fragmentation bombs.

The Chinese soon found that if a hollow tube closed at one end is filled with gunpowder and ignited at the open end, fire and hot gases would come rushing out, and that the tube itself, if not secured, would be propelled in the opposite direction. This discovery led to the invention of **rockets**, devices like Roman candles—and, ultimately, cannons and guns. One early application involved attaching these powder-filled tubes to arrows (just behind the tip). The resulting rocket-arrows rendered bows obsolete. They were launched against adversaries in large numbers. Another weapon, the "**fire-lance**," looks rather like a larger version of the rocket arrow, but the open end of the two-foot-long tube of gunpowder points toward the tip of the spear so that it functions as a flame-thrower (see Figure 8-1 below). Fire-lances were especially effective in defending city walls against attackers. They became increasingly formidable weapons as bits of metal, shards of broken porcelain, arrows, and stone or metal pellets were embedded in the gunpowder so that they came spewing forth when the device was ignited, pelting those targeted.

As the Chinese developed more powerful gunpowder, they replaced the bamboo tubes initially used with bronze or iron ones capable of withstanding the pressure generated. They also learned that using projectiles that more nearly matched in size the inside diameter of the tubes resulted in greater missile velocities and increased range of the weapon. By this time, of course, these **proto-guns** or **proto-cannons** (in the case of the larger ones) were no longer mounted on spears and no longer used as flame-throwers. Their function was to hurl a projectile as fast and as far as possible. Given that gunpowder was now being made that would explode, only two small steps had to be taken in order to produce a true gun: (i) the size of the shot had to be closely matched to the inside diameter of the tube (to maximize the effect of the gunpowder exploding behind it), and (ii) a method of igniting that explosive charge behind the shot had to be devised. This was usually accomplished by means of a small "touch-hole" through the side of the tube or barrel near the butt end. The earliest example of such a weapon recovered in China dates from the late thirteenth century. Hundreds of artillery pieces from the mid-fourteenth century survive. By that time cannon had assumed a place in China among the many instruments of war.

Because of the vast differences—practical, cultural, symbolic, etc.—between pyrotechnics and firearms, one might be inclined to think

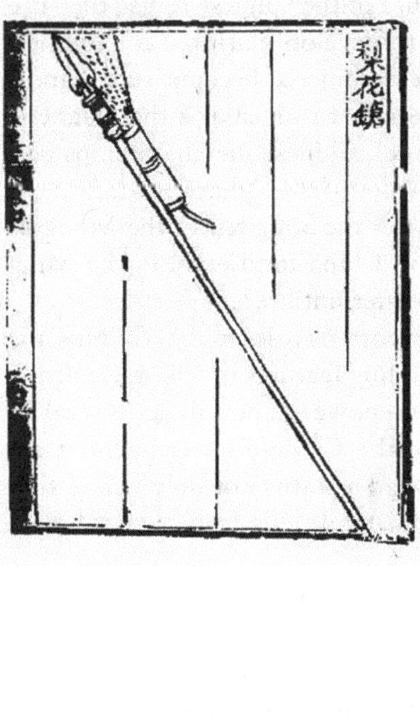

Figure 8-1. Fire-Lance. Illustration from a fourteenth-century (Ming dynasty) Chinese military treatise on gunpowder weapons, the *Huolongjing* ("Fire Dragon Manual"), compiled by Jiao Yu and Liu Bowen.

that gunpowder assumed a third identity when it was put to use in warfare, an identity distinct from "the fire drug that makes pyrotechnics possible." But given that the development of military uses of the fire drug in China and its pyrotechnical functions are both part of one on-going, step-by-step historical process, it would be misleading to say that gunpowder assumed a third identity there when it was first put to use in warfare. Rather, **different uses** were made of the **same technology.** On the other hand, to the extent that distinct types of fire drugs were developed to serve different purposes (such as more powerful weaponry and more colorful pyrotechnics), then it could be said that **distinguishable technologies with separate identities** had come into existence. Admittedly, the distinction between (i) different uses of the same technology and (ii) the utilization of different technologies is not a sharp one (more on this matter in Section D below).

Although military tactics and operations in China were adjusted over the centuries to take into account the new weapons as they emerged, the overall effects were far from revolutionary, either militarily or socially. Part of the reason that no such rapid fundamental changes took place is the time it took for a sufficiently powerful form of gunpowder to be developed in the step-by-halting-step process that led to it. In Europe, on the other hand, gunpowder suddenly appeared as a powerful explosive—which is, in itself, all-but-conclusive proof that it was not invented there. But, one wonders, why in subsequent centuries was there no Chinese analogue of the role that gunpowder—and other inventions, many of them of Chinese origin—played in Europe's ascension to a position of wealth and power? This is a complex question that has been the subject of much speculation and is still being debated. However, some things are clear. The answer to the puzzle has nothing to do with the supposed failure of the Chinese to use the "fire drug" for military purposes (as seen above) or to develop maritime capabilities. During the Song dynasty (960–1279) seaborne commerce became very important. Making use of the fore-and-aft rig, the sternpost rudder, and the magnetic compass (crucial to later developments in Europe), Chinese merchant ships carried goods to destinations to the east and south—out of the sight of land. And the Song built a navy to protect the trade routes. After the Song fell to the Mongols in 1279, such enterprises were not to be seen in China until early in the Ming dynasty (1368–1644, see below) and never thereafter until recent decades.

Three types of factors which surely played important roles in determining the course that events took in China are (i) long-standing features of Chinese culture, (ii) China's experience with foreigners and foreign powers, and (iii) actions taken by powerful groups within the country or by the Chinese government itself. Traditionally, merchants and artisans in China had a status not only below that of scholars but also below that of peasants. But while this feature of Chinese culture may have influenced the course of events at one juncture or another, it cannot be of much help in accounting for the difference between the inventive and enterprising Song period and the increasingly insular and moribund era that began in the fifteenth century. Another factor, often cited as primarily responsible for this change of course, was the Chinese civil service. Until 1949 the country was ruled by a complex administrative system staffed by salaried bureaucrats selected by examination and promoted by merit. The system had roots going back to the Q'in and Han dynasties (to the late third century BCE) but assumed the central role it came to play only in the Tang dynasty (618–906). Faced with the threat to the imperial government posed by the powerful nobles of the aristocratic warrior class, the first Tang emperor sought to limit their power by staffing the administrative offices of the government with civil servants selected from the literate gentry on the basis of examinations. Difficult by design, these civil service examinations became the gateway to the good life for generations of Chinese.

As time passed, the examinations became more elaborate and more narrowly focused on classic Confucian texts. Preparing for them became the primary function of education, which stressed memorization and familiarity with texts. A government bureaucracy staffed with scholar-officials thus educated could hardly fail to exert a conservative and traditionalistic influence upon the country. While some writers have seen this feature of Chinese culture as antithetical to innovation and exploration, such certainly was not always the case. Indeed, the improved administration achieved by means of the civil service program played a vital role in the overall Tang achievement, which included land reform, improved transportation, greatly expanded foreign trade, a richer diet, new forms of artistic and literary expression, etc. Similarly, the Song dynasty was a period of technological innovation, cultural achievement, and (as noted above) contact and trade with distant lands. Thus, while the civil service did attract talented individuals, preparations for the examinations did absorb time and resources, and the educational system did foster traditionalism, other factors conspired to depress invention and entrepreneurship in China.

Especially important was the arrival in China of foreigners—whether to pillage, subdue, or trade—and the reaction thereto. The Song proved incapable of defending China against the fierce, nomadic people who pressed in from the northeast and northwest. Eventually, the Mongols, led at first by Genghis Khan and later by Kublai Khan, overran China. There followed a period of almost a hundred years of Mongol rule, during which garrisons were stationed in Chinese cities and administrative positions were often staffed with foreigners. Ineffective rulers followed, and resentment and discontent eventually blossomed into full-scale rebellion, which brought an end to Mongol rule in China by 1368. The Chinese then set about re-establishing the civil service, rebuilding the country's defenses, restoring the canals and irrigation systems, etc. Beginning with the Mongol occupation and intensifying thereafter, an antipathy toward foreigners and things foreign took root in China. It was reinforced by the prosperity and stability that ensued under Ming rule, which helped to foster an attitude of superiority regarding Chinese ways and institutions and an indifference to all things foreign.

Even so, marine commerce flourished during the earlier part of the Ming dynasty. Most remarkably, between 1405 and 1433, the imperial court mounted a series of large naval expeditions involving dozens of massive ships and tens of thousands of men. The purposes of these voyages were complex and are not now well understood. Certainly they were intended to convey a message of power and grandeur to the smaller kingdoms and territories to the east and south. Although there clearly was an intent to establish a sphere of influence, rulers willing to accept Chinese suzerainty received lavish gifts from the Ming emperor rather than be forced to pay large sums in tribute. The expectation was that they, in turn bearing gifts, would journey to Beijing to pay respects to the emperor. Some who would not accept the emperor as overlord were replaced by less recalcitrant leaders. And although there was an economic aspect to the expeditions, trade was not their primary mission: exploration and discovery must be reckoned among the reasons for these voyages. Small squadrons and single vessels were dispatched to investigate obscure islands and harbors. Not only were Indonesia and the Philippines explored, but the Chinese sailed over to Sri Lanka and India, then to the Persian Gulf region, the Red Sea, and down the African coast.

As remarkable as these voyages were, even more striking was their abrupt cessation. Ironically, their demise was rooted in the conditions under which they were first launched. The financial backing for the voyages came from funds under the direct control of the emperor, not from the usual government revenues administered by the civil service. Partly as the result of the rivalry between those closely associated with the imperial court and the civil servants and partly because of their traditionalistic and insular perspective, the civil servants had opposed the voyages from the very beginning. Thus, when a new young emperor, who lacked the financial resources of his predecessors, sought funding from the national treasury to continue the voyages, the government bureaucracy used delaying tactics and obstructionism to derail the project. Subsequently, the Chinese navy was confined to China's coastal waters and allowed to fall into decay—to the extent that it could not protect Chinese maritime activity (by then also restricted by imperial edict to coastal waters) from piracy. In time, Chinese citizens were forbidden to travel abroad, and the country settled into an isolated and relatively static pattern of existence.

2. The Impact of Gunpowder in the West

Remarkably, in the very same century in which Chinese sailors had crossed the Indian Ocean and sailed down the African coast, Portuguese mariners rounded the Cape of Good Hope and by 1498 had made their way across to India from the opposite direction. Soon thereafter (in 1516), Portuguese ships arrived in Guangzhou (Canton)—during the very period that the Chinese were pulling back to their home territory and limiting contacts with foreigners. The Chinese government did permit trade with the Portuguese, and other Europeans when they arrived, but maintained strict controls over it and limited it to a very few ports. Ironically, these and other voyages of exploration and entrepreneurship undertaken by the Western Europeans during this period would not have been possible without the sternpost rudder and magnetic compass, knowledge of which came from the Chinese, and the lateen sail (fore-and-aft rig), which also may have come from China by way of the Arabs. Also essential to the European dominance of the world scene which began to unfold in the fifteenth century was another Chinese invention: gunpowder. And unlike the course of events in China, the introduction of gunpowder and firearms into Europe was followed by radical changes, not only in warfare but also in many other areas as well. In most basic terms, the primary difference between the two situations is that exploration, invention, and entrepreneurship were strongly encouraged and rewarded in Europe at the time but, by and large, had ceased to be so in China.

Perhaps the most obvious and impressive effect of the introduction of gunpowder into Europe was to render obsolete the high walls and soaring round towers that had, until then, rendered fortified towns, cities, and castles so nearly impregnable that prolonged sieges were the only effective means of subduing them—by starving out the defenders! Once firearms appeared in Europe, weapons development proceeded at a rapid pace. The first record of **cannon** being in Europe dates from 1326. However, these early "bombards," as they were called, were very inaccurate, often blew up, and did more to terrorize a besieged population than to inflict actual physical harm. Even so, by the closing decades of the Hundred Years' War, which ended in 1453, cannons were being made that were sufficiently powerful and reliable to batter down thick stone walls and to subject the towns and cities they protected to bombardment (see Figure 8-2 below). Consequently, fortified places that previously might have held out for months under siege capitulated in a few days. Most notably, Constantinople, which was defended by massive double-walled fortifications that had withstood twenty sieges over the centuries, fell to the Ottoman Turks after slightly less than seven weeks of bombardment that same year (1453). The centerpiece of this operation was an enormous cannon

Figure 8-2. Early Cannons. These cannons were made and used during the Hundred Years' War. They were abandoned at Mont Saint-Michel in 1434. Greenshed. (https://commons.wikimedia.org/wiki/User:Greenshed)

twenty-six feet long that was able to hurl a stone ball weighing half a ton. It was cast in a makeshift foundry on site under the direction of a Hungarian metallurgist and gunpowder technician, one of the most skilled of the time.

Perhaps most significant in historical terms was the essential role firearms played in the endeavors of Portugal, Spain, England, the Netherlands, and France as they fanned out around the globe, carving out colonial and commercial empires. As noted above, gunpowder arrived in Europe during a period in which innovation, entrepreneurship, and exploration were encouraged and rewarded. By this time the Medieval Industrial Revolution was in full flower, and the benefits of increased productivity (made possible by new or improved technologies) and trade were very much in evidence. There was also a growing fascination with foreign lands and the treasures they might hold. The desire for economic gain thus was reinforced by the allure of exploration and discovery. As might be expected in such a context, merchant-adventurers led the way out into the wider world. Probably the most well-known of these were members of the Polo family of Venice, whose travels to China in the thirteenth century were made famous by a book written by one of them, Marco Polo.

Another group of fortune-seeking travelers were the mariners, who fanned out across the oceans to establish trading stations, bases of operation, and colonies around the globe. Of course such ventures depended upon financial backing as well as sailors (as seen above in the case of Ming-dynasty China), and the most likely sources of the large sums required were potentates or governments. Accordingly, anyone with a keen interest in marine navigation and exploration who had direct access to funding sources of this kind likely played a significant role in the developments under discussion. Such a man was Prince Henry ("the navigator") of Portugal (1394–1460) who probably did more than anyone else to speed the process along. He devoted nearly all of his adult life to assembling the best possible resources for marine navigation, including ships fitted out with fore-and-aft rig (triangular sails), and he launched expedition after expedition down the west coast of Africa. Each of these voyages was an economic venture as well as an exploratory expedition, the hope being that ships would return with goods of various sorts, obtained in trade or otherwise, sufficiently valuable to defray the costs incurred and even turn a profit.

Unfortunately, it became clear right away that slaves were one of the most profitable commodities that could be brought back to Portugal from the West African coast. And while, in their early expeditions, the Portuguese did attempt on their own to capture African people for enslavement, they soon found it much more advantageous to obtain enslaved peoples in trade from Arab dealers or from other Africans. In the centuries that followed, the slave trade and the use of slave labor were important parts of the far-flung commercial activities not only of the Portuguese but also of the Spanish, French, Dutch, and English. What is noteworthy is not that slave holding and trading were practiced by these Europeans as they set out to explore the world beyond the horizon and to enrich themselves in the process. At the time, slavery was to be found more or less everywhere. What is remarkable—and horrific—is the *scale* of the European operations: millions of slaves were transported enormous distances, many of them across thousands of miles of open ocean.

Despite Prince Henry's resources and perseverance, it took a long time to overcome the obstacles, human and otherwise, standing in the way of long-distance navigation. When he died in 1460, Portuguese mariners had pushed only as far south as present-day Sierra Leone. Even so, by this time the profitability of maritime expeditions to Guinea had been amply demonstrated, navigational techniques had been improved, and the reluctance of sailors to venture very far into uncharted waters had been somewhat overcome. Thus, even though further exploration and trade were in the hands of private entrepreneurs for the next few years, Portuguese mariners pushed farther east and south as far as present-day Gabon, perhaps crossing the equator. Then in 1481 a young man, who shared Prince Henry's enthusiasm for exploration and who dreamed of breaking the Arabs' monopoly on the lucrative trade with the East, ascended the Portuguese throne as John II (1455–95). He hoped to show that it was possible to sail around the southern tip of Africa and reach India and China by sea. At the time it was not known whether the Indian Ocean was like the Mediterranean Sea, with land to its south connecting to Africa, or whether Africa had a southern terminus, with the Indian and Atlantic oceans merging to its south. Accordingly, John II immediately set about establishing or rebuilding fortified outposts along the West African coast, each to serve as a trading center and as a base of operations for further exploration. Thus, in rapid succession, he dispatched a series of expeditions farther and farther down the West African coast, culminating in Bartolomeu Dias rounding the southern tip of Africa in 1488.

The sea route to India lay open; but before the Portuguese could take advantage of it, they were confronted with a new challenge from an old adversary. Unable to convince King John and his advisers of the feasibility of reaching China and the fabled Spice Islands by sailing west from Portugal, the Genoese mariner and visionary, Christopher Columbus (1451–1506), sought support for his scheme from the Spanish crown. Here too he was unsuccessful at first but eventually received the backing he required. Columbus's case rested on his conviction that the Earth was much smaller than it actually is and, indeed, much smaller than it was judged to be by contemporary experts on the basis of evidence available at the time. Sailing west in 1492, Columbus did make landfall some 4,000 miles west-southwest of Spain; however, the islands he claimed for Spain did not lie off the coast of Asia, as he continued to believe the rest of his life. In any case, by sponsoring Columbus's voyage Spain had, in a single stroke, embarked upon the road to exploration and empire-building on a global scale. Not surprisingly, Portugal regarded Spain's initiating such activities as an intrusion upon its "turf." Hostilities threatened, but negotiations led to the Treaty of Tordesillas (1494), which assigned separate zones of exploration to the two kingdoms.

Of course, this agreement did not prevent other countries from joining the search for new trade routes and sources of valuable commodities, especially if this could be done in areas far from the centers of Spanish power and resources, i.e., far to the north or far to the south. It is hardly surprising that England, an island nation already deeply involved in maritime commerce, was the next to enter the competition. Indeed, Columbus's brother, Bartholomew (c.1461–1514), had tried unsuccessfully to persuade Henry VII of England (1457–1509) to sponsor Christopher's initial voyage west. In 1497 another Italian mariner, John Cabot (c. 1450–c. 1499), did sail west across the Atlantic under the English flag. His voyage, which provided the foundation for English colonization in North America, was financed by a group of Bristol merchants and authorized by Henry VII. In the ensuing decades mariners from Spain, Portugal, and England explored what are now known as the east coasts of the Americas. During this period the Europeans gradually came to realize that they were confronted with a whole new world rather than Asia or its environs.

Spain quickly established self-sustaining colonies on Hispaniola (present-day Haiti and the Dominican Republic) and Cuba, with Spanish settlers and enslaved Indigenous peoples engaged in growing crops, raising livestock, and mining. Then, in short order, Spanish forces numbering only in the hundreds marched into Mexico (1519) and Peru (1531). Although confronted with many thousands of hostile warriors, the Spanish, through luck, treachery, and superior weaponry (steel swords and pikes, steel helmets and armor, firearms, horses, etc.), quickly took possession of the Indigenous Mexican and Peruvian cities and helped themselves to everything of value: gold, silver, precious stones, etc. In the process, they wreaked death and destruction upon everything in their path.

Although gunpowder played an important role in the Spanish conquests in the Americas and in the Portuguese exploration and trading activities along the West African coast, the two countries might have succeeded at these enterprises even without firearms, since the weapons of the inhabitants of those lands (with the possible exception of the Arab slave traders) would still have been no match for those of the Europeans. But when the Portuguese mariner, Vasco da Gama (c. 1460–1524), rounded the southern tip of Africa in 1497 and headed up the east coast of Africa on a mission to establish trade relations with India and the East, he confronted a fundamentally different state of affairs: a pre-existing network of maritime commerce being carried out by the Arabs, who certainly had the use of gunpowder and may have been the source of the Europeans' knowledge of it. This network included trade between India and ports along the east coast of Africa as well as between India and the eastern Mediterranean region by way of the Red Sea and the Persian Gulf. Da Gama was face-to-face with the very trading operation that the Portuguese had set out to outflank!

In order to protect their monopoly, the Arabs did their best to thwart da Gama's mission by preventing him from receiving any assistance or cooperation.

Despite these obstacles, however, da Gama managed to sail across the Indian Ocean to the port of Kozhikode on the west coast of India. But neither the ruler nor the inhabitants of Kozhikode were impressed with the goods da Gama brought as gifts or trade items, and he returned home with only samples of Indian wares. A follow-up expedition in 1500 tried, without notable success, to negotiate with the sheiks along the East African coast, but it was able to launch a trading operation in Kozhikode, only to have it attacked by the Arabs, resulting in much loss of life. Reprisals by the Portuguese followed, and in ensuing years they dispatched a series of ever larger and more powerful fleets to the Indian Ocean for the express purpose of replacing the Arab trade monopoly with a Portuguese one. Matters came to a head in 1509 when, with support from the Ottoman Turks, a coalition consisting of Indian states, Egypt, and Venice, all of which had been profiting handsomely from the established trade network, assembled a fleet to challenge the interlopers. Although the coalition forces had nothing in their arsenal comparable to the Portuguese ships, built strong enough to withstand the storms of the South Atlantic and bristling with cannon from stem to stern, a fierce battle ensued near Diu on the coast of Gujarat. Outmatched, the coalition forces finally broke off the engagement. Thereafter Portuguese control of marine commerce in the Indian Ocean remained uncontested until other European colonial powers entered the picture.

The Portuguese proceeded to strengthen and extend their network of secured centers of operational support and trade. Some were established through negotiated agreements with local authorities (e.g., Macao in China), while others were acquired by military conquest, most notably Goa (on the west coast of India) and Malacca (on the strait separating Sumatra from the Malay Peninsula), the most important trading center in the East Indies. Around 1516 the Portuguese established a trading post in the Moluccas, the Spice Islands themselves.

A new era in European exploration and colonization began about 1600, and England, France, and the Netherlands soon became major players. By and large, instead of ventures into unknown lands and seas in search of riches, competition and armed conflict between the European colonial powers became the order of the day. In many cases bloody struggles in colonial areas were projections of hostilities in Europe.

3. Confrontation of the Elite Swordfighter with the Gun in Premodern Japan

Not only did gunpowder usher in fundamental changes in the very nature of warfare in Europe and enable a few European nations to fashion globe-spanning empires, but the general availability of firearms placed the continued existence of elite warrior classes in extreme jeopardy. The strength, skill, and courage of a swordsfighter or mounted knight are of little use against a peasant armed with a gun. Access to **firearms** is thus the key issue, and that, in turn, is a matter of time and conditions. This point is aptly illustrated by developments in Japan between the 1540s and the 1860s. The fearsome samurai warriors of Japan were able to retain their privileged place in society well into the modern era while keeping the sword as their weapon of choice only because of the confluence of some unusual conditions. Among the most significant factors were the policies followed by the rulers of Japan, beginning with Toyotomi Hideyoshi (1537–98), as they strove to maintain control of what had been a very fragmented land. Having unified the country under one government by about 1590, Hideyoshi set about establishing laws contrived not only to prevent the formation of any force capable of challenging the central government but also to stop the disruptive changes then taking place in the country, which he viewed as a threat to its stability. Peasants and farmers, who made up 80 percent of the population, as well as artisans and merchants were forbidden to possess weapons of any kind, especially the swords associated with the skillful combat of the samurai class.

A follow-up edict three years later reflects what seems to have been the guiding principle of many of the policies of the Tokugawa shogunate, which came to power after Hideyoshi and governed Japan for the next two and a half centuries: the way to achieve political stability is to prevent society from changing. This edict put an end to social mobility and solidified the existing class structure, with warriors at the top, followed by farmers and peasants; below them came artisans and, lowest of all, merchants. Peasants were forbidden to leave farming to take up a trade or to work as hired laborers. Military men could not seek employment in agriculture or the trades or leave one master for another without permission. In this hierarchical society, the welfare and interests of those of higher rank was paramount and the loyalty of those of inferior rank was demanded.

To limit the power of any possible challenger, Tokugawa Ieyasu (1543–1616), who seized control of the country by military means after Hideyoshi's death and assumed the title of Shogun in 1603, divided Japan into 250 administrative units. The feudal lords (daimyo) who presided over each of these units reported directly to the shogun in Edo (later renamed Tokyo). The daimyo and their families were required to reside there during part of the year. When the daimyo returned to their fiefdoms, they had to leave some family members in Edo—in effect, as hostages.

In another move to secure its position, the shogunate (in 1607) set about gaining control of the production of firearms. After requiring gunsmiths to move to a centralized location in Nagahama, where they could be monitored, the government appointed a commissioner of guns. No order for firearms could be filled without his approval. In practice he approved only orders from the central government. Although this scheme was not fully implemented right away, in due course the resulting government monopoly became well established. With the country unified, tightly controlled, and facing no external military threats, the government had little need for guns; hence its orders gradually decreased and firearms faded from the scene. And, as noted in the Introduction, they were not to reappear in significant numbers for 250 years.

That is not to say that guns disappeared entirely from Japan during this period. Indeed, recent research suggests that guns were more common in Japanese villages during the Tokugawa period than might be supposed. But importantly, their usage was restricted to hunting, with harsh consequences for employing firearms for combat. A 1610 edict, for example, decreed that farmers using firearms to fight over land and water could have their entire villages destroyed. However, this only further demonstrates the profound effect that society can exert on technological adoption. Despite the power of firearms technology in combat, social factors, including the preservation of the warrior culture of the samurai class and the prevention of armed uprisings by peasants, resulted in a period during which firearms were used by peasants for hunting but were not generally used in combat.

Far from being an isolated program, the decrees that led to the virtual elimination of firearms as weapons were part of a concerted effort to prevent Japanese society from changing. The justification for many of the policies followed during this period was the belief that foreign influences posed a threat to those in power and to the stability of the nation. Accordingly, much energy was focused upon preventing foreign influences of any sort from infiltrating Japanese culture and disrupting traditional patterns of living. Simply put, traditional Japanese ways and views were championed as good and desirable, while all things foreign were condemned as evil and corrupting.

The first group to feel the full force of this growing conviction was the Japanese Christians. The Portuguese traders, who arrived in Japan in the 1540s and had introduced firearms into the country, brought Jesuit missionaries with them. These priests were remarkably successful in making converts, especially in the region around Nagasaki where most of the trading took place. The vast majority of these converts were farmers and peasants, but some of them were persons of higher rank, including daimyo (feudal lords) and ladies of the court. While there were no major problems at first, the arrival of more and more traders of different nationalities accompanied by clerics of different affiliations and the resulting maneuvering for advantage led to involvement in Japanese politics. As time passed, occasions arose in which the Christians were—or were perceived to be—a threat to Japanese rulers' hold on power and to the stability of the country. At the same time, there was an increasingly widespread fear of Christian and European domination. As a result, both Hideyoshi and his successor, Tokugawa Ieyasu, issued decrees banning Christianity from Japan and ordering the departure of the missionaries.

While Hideyoshi's edict was not rigorously enforced, especially at first, it marks the beginning of the persecution of Japanese Christians, which, while intermittent, grew more intense as time passed. Churches were destroyed, Japanese Christians exiled or tortured, and missionaries expelled. Beginning about a year after Ieyasu's death in 1616, large numbers of Japanese Christians who refused to recant were beheaded or burned at the stake along with some of the missionaries. The Spanish traders were expelled in 1624 along with the monks they had brought with them from Manila. Finally, in 1638, in what may be viewed as a peasant rebellion or as an act of desperation by an oppressed group, 37,000 Christian peasants led by a handful of samurai converts holed up in an old castle on the Shimabara peninsula near Nagasaki and held out for three months against some 100,000 government troops. In the end all (or nearly all) of them were slaughtered, effectively bringing an end to Christianity in feudal Japan.

One might think that the eradication of Christianity would have allayed fears of foreign domination and political and social disruption. Certainly, it had been the intent of Ieyasu to address such threats when he issued his 1614 edict outlawing Christianity. Even so, he welcomed foreign trade and vigorously promoted the development of the Japanese merchant marine and the establishment of Japanese trading posts abroad. He had even made an English mariner, Will Adams, who arrived on a typhoon-battered Dutch ship, his adviser on shipbuilding and navigation. Instead, the Shimabara incident was viewed as confirming the dangers posed by foreign influences. Following it, the Portuguese traders were ordered to depart. (The British traders had already left of their own accord.) In the end, there remained only a small contingent of Dutch traders who were confined to a tiny island in Nagasaki harbor.

The depth of the shogunate's commitment to shielding Japan from foreign influences is demonstrated by its having ordered the cessation of all overseas trading operations being carried on by its own citizens. No Japanese citizen was allowed to go abroad thereafter, and those who had been out of the country for five years were not allowed to return. Since by this time the Japanese had developed an extensive overseas trading network along the Chinese and Southeast Asian coasts and in the Philippines, thousands of Japanese were left stranded abroad by this edict. To make certain that the ban on travel outside the country was not violated, the shogunate decreed that no large sea-going ships were to be built, only fishing boats and small vessels for transporting goods between the Japanese islands. As a security measure, even travel from the fiefdom of one daimyo to that of another was severely restricted and carefully monitored.

In addition to the policies of the government and the growing antipathy toward all things foreign, two other major factors contributed to the reversion to edged weapons. One was geopolitical: the location and insular nature of the country together with the reputation of the Japanese as fierce fighters made Japan an unlikely target for invasion. Martial values emphasizing prowess-at-arms, which permeated Japanese culture at the time, was the other. In this context the sword had tremendous symbolic value closely tied to a samurai's honor and identity. Even having a family name (surname) depended on the right to carry a sword, and only samurai had that right. That a gun enabled a humble peasant to kill a samurai was—just unthinkable, outrageous! Sword-making and swordfighting techniques, like so many technologies, provide an opportunity for aesthetic expression. Both were highly developed art forms in Japan long before the arrival of the Europeans. But neither the matchlock gun nor the bodily movements involved in loading and firing it fit these patterns. Perhaps most importantly, the use of firearms changes the way battles are fought, greatly reducing the incidence of individual hand-to-hand combat and thereby depriving warriors of an opportunity to demonstrate their strength, skill, and courage. It is difficult to overestimate the significance of this effect in a culture in which military glory was coveted above all else by essentially all men of inherited high status. In such a context, it is no surprise that, given a choice, these warriors preferred to fight with swords rather than guns.

4. Later Developments in the West

In China and Europe, on the other hand, there was no comparable confrontation between firearms and elite warriors. In both cases, aristocratic warrior classes were no longer in the picture when firearms made their appearance. In China, as seen above, the first Tang emperor had devised a strategy for limiting the power of the military aristocracy, with the result that China came to be ruled by a cadre of scholar-officials. The absence of a class of elite warriors in China at the time firearms became significant probably is one of the reasons that the invention of gunpowder was not followed by revolutionary changes in warfare or in society. The mounted, armored knights of Europe would have confronted a challenge comparable to that faced by the samurai warriors had not their power monopoly already been broken by the Swiss pike square or pike phalanx. This highly maneuverable, dense formation of foot soldiers defended itself with wooden poles eighteen feet long tipped with four-foot steel spikes. A horse charging into a wall of these spikes would be impaled and his fallen rider quickly put out of action. Indeed, for a hundred years or so the pike square was the most formidable fighting machine on the European continent. However, over time improvements in firearms and their use eroded this advantage. Following the introduction of the **ring bayonet** in 1690, the pikeman disappeared altogether, and musket-armed infantry made up the core of European armies.

The very antithesis of the elite warrior class made its appearance in the armies of revolutionary France. Finding itself under attack and lacking a trained officer corps—as aristocrats, they had nearly all either fled the country or been executed—the French government set about mobilizing the entire nation and all its resources to meet the challenge. The centerpiece of this effort was the conscription of young men in enormous numbers. During one period, some six hundred thousand to one million men were massed on the frontiers of the country. Going into battle, the French infantry formed up in huge squares and rolled over the opposing force by the sheer weight of numbers. The rest of the population was pressed into supporting the war effort. In an endeavor to produce one thousand gun barrels a day, some two hundred forges were set up outdoors in Paris.

C. The Impact of Industrialization on Women and the Family

The mechanization of manufacturing processes that took place in the late eighteenth and early nineteenth centuries did more than obliterate the economic value of the skills of traditional craftspeople, change the character of work performed, and effect a separation of workplace and dwelling. It set in motion a series of developments that brought fundamental changes in the nature and distribution of the contributions made by family members toward meeting their collective needs. But these did not all come about right away. One of the more radical changes was the eventual elimination of the economic contribution of (non-adult) children. As noted in Chapter V, an important feature of the agriculture revolution was that children became a major economic asset. And as economic activities developed, they played a similar role in cottage industries and family-operated shops in commercial centers. Up to that point, children working along with and learning from adult family members would hardly have seemed remarkable or a cause for concern. But as the Industrial Revolution proceeded and even very young children were put to work in mines, factories, and textile mills, outrage at children being separated from their families and working long hours in unsafe and unhealthy conditions led to the enactment of child-labor laws. (The advantages believed to accrue to an educated citizenry and workforce were also factors.) Thus, one effect of industrialization and child-labor laws was to eliminate the positive economic role which children had played since human beings took up the practice of agriculture some 10,000 years ago.

There was also a reconfiguration of the household tasks performed by husbands and wives. A significant number of tasks that typically had been men's responsibilities ceased to be carried out at the home-site. For example, prior to the Industrial Revolution (when the vast majority of people lived on farms or in villages or small towns), men cut, hauled, and chopped up firewood for cooking and heating. As industrialization and urbanization proceeded, these functions increasingly were taken over by iron stoves fueled with coal delivered to peoples' homes—both paid for by cash earned in wages. Similarly, men had once fashioned numerous household items from wood (simple pieces of furniture, spoons and ladles, bowls and platters, etc.) and had made or repaired leather articles (belts, shoes, harness, etc.); but with industrialization the place of these hand-crafted items was taken by manufactured products: tin-ware, earthenware, enamelware, factory-made furniture and shoes, etc. Furthermore, as slaughtering and butchering operations became concentrated in large meat-packing plants and refrigerated transport became available, most men ceased to take part in this work as well. Thus, as industrialization proceeded, men's primary economic role became that of wage-earner.

In contrast, tasks that traditionally had been the responsibility of women changed little, more slowly, or only partially during the early decades of the Industrial Revolution. Women continued to care for infants and small children, to prepare meals, to do the cleaning, washing, sewing, and mending, to care for the sick, to plant and tend the kitchen garden, and so on. Furthermore, in poorer families (who made up the majority of the population), married women, in order to provide their families with the barest essentials, often found it necessary to supplement what their husbands could provide by taking in roomers and boarders, by doing laundry (at their own homes or at another's), by working as cooks or maids or laborers in factories, etc. Readers will find it difficult to imagine living under the miserable and desperate conditions endured by most people during this period. Evidently, whatever relief industrialization would bring had yet to arrive. One never-ending chore that industrialization did promptly remove from women's daily routine was spinning thread. Thus ended a pattern of activity going back twenty or thirty thousand years so closely associated with women that the word "distaff" (a tool used in spinning) was once commonly used to refer to women and to women's work or concerns.

Ironically, however, power-driven looms and spinning machines did not lighten women's heavy workload. What mechanized fabric production did do was to make woven fabric available at a reduced price and in much greater quantities. Furthermore, the invention of the **cotton gin** around the turn of the nineteenth century (see Figure 8-3 below) transformed cotton from a luxury fiber into a utilitarian one. An important quality of cotton cloth is that, unlike linen or wool (which it replaced for many purposes), cotton can be washed fairly easily. The net result of these developments was that there was a lot more laundry to be done, and doing the laundry became a burdensome weekly chore, for which earlier centuries offer no clear precedent.

In industrialized countries, the twentieth century brought major changes to the domestic scene, most related to the availability of current electricity in the home. Not only did electricity eliminate the need to clean the chimneys of lamps, to trim their wicks, and to replenish their oil, but it provided the motive power for an entire array of household machines that enabled their users to achieve better results and to do so in less time. **Washing machines, sewing machines**, and **vacuum cleaners** powered by **electric motors** were introduced early in the century, and as mass-production methods brought their prices down, they became available to a larger and larger segment of the population. Another home appliance which made an even more rapid transition from technical solution to mass-market product was the refrigerator. Once a suitable refrigerant had been found (Freon, later determined to be a threat to the ozone layer) and a satisfactory pump design developed, the electric refrigerator went into production, initially as a high-end, specialty product; but in only fifteen years or so it had become a moderately priced, mass-market commodity! This pattern, in which a new device, initially marketed at a high price to a limited number of customers, is developed into a mass-produced article offered at a fraction of the original cost, is characteristic of the way new products reach the bulk of the population in a market economy. Among many examples are automobiles, televisions, calculators, computers, microwaves, and cellphones.

Other popular electrical devices introduced early in the twentieth century made use of resistance coils to produce heat. The **electric iron** was a major improvement in that it was lighter, maintained a constant temperature, and did not require that a fire be maintained to heat it, which was a special advantage during the hot summer months. **Gas-fired cook-stoves**, which offered similar advantages (except for the matter of weight), came into widespread use during the early decades of the twentieth century. Then in the 1920s electromagnetic thermostats were perfected so that by 1930 electric ranges could be found in many kitchens, especially where natural gas was not available. In addition,

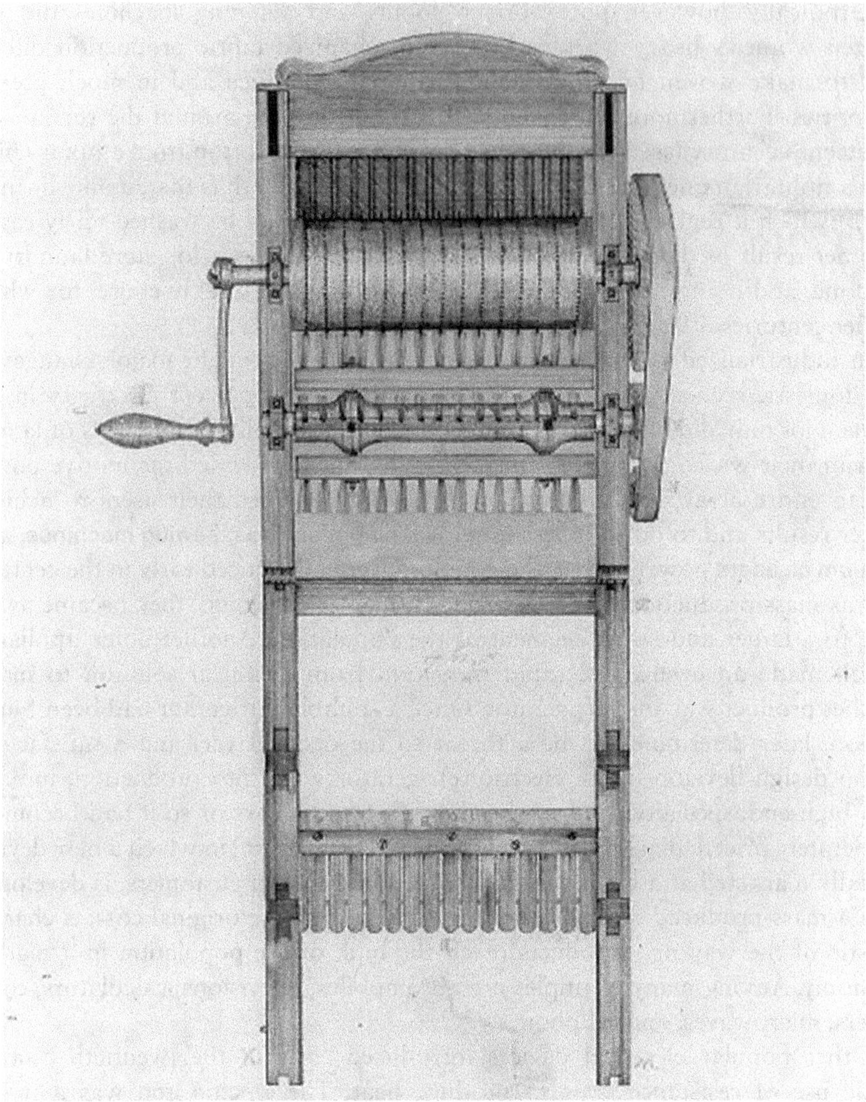

Figure 8-3. The Cotton Gin. Eli Whitney (1765–1825) invented the cotton gin (short for 'engine') in 1793, making it feasible to process cotton at an unprecedented rate. The social and political effects of this piece of technology, beyond creating a greater laundry burden to be borne by women, included rejuvenating the southern United States' plantation economy, creating greater demand for slave labor and thereby contributing to the social, economic, and political conditions leading to the US Civil War. This drawing was made by the United States Patent Office.

other electrically powered machines were marching to the aid of the beleaguered homemaker: the **automatic washing machine**, the **electric dishwasher**, and the **clothes-dryer**; but their arrival in significant numbers did not come until after the Second World War. Not to be forgotten in this context are the advances in medicine and healthcare, which greatly reduced the amount of time that women spent nursing the sick at home.

With all the labor-saving devices available to the typical American housewife of the 1960s, 1970s, and 1980s, one might be surprised to learn that studies show that these women logged about as many hours per week doing housework (or running errands connected with it) as had their mothers and grandmothers in the earlier decades of the century. How can this be? Many factors are involved, including major changes in the tasks performed by mothers and housewives and the restructuring of a number of retail and service businesses. In the case of the more affluent, the availability of the new machines and products meant that women were doing more of the cooking, cleaning, and laundering chores themselves, with less hired help in the home and less support delivered to their door by commercial enterprises. For the bulk of the population, the new domestic regime brought relief from drudgery, a higher standard of living, and a level of health and cleanliness unlike anything that had existed previously anywhere. At the same time, there were major declines in domestic employment, commercial laundries, delivery services of all sorts, and house-calls by physicians. Along with these changes, there appeared a new clause in the job-description of "housewife": chauffeur and motorized gofer.

At mid-century, many women were already working outside the home—as nurses, teachers, secretaries, retail clerks, telephone operators, textile workers, etc. In ensuing decades, not only was an increasingly larger fraction of the female population employed, but more women were working in industry and pursuing careers in the professions: medicine, law, accounting, business and public administration, engineering, etc. As this process unfolded, women had fewer children, reversing a trend that accompanied the advent of the agricultural way of life. But this watershed in human affairs was not primarily the result of technological innovation. Rather, that responsibility lies with social, political, and economic developments. Even so, new products, techniques, and devices played an important role. An oft-cited example is the ready availability of inexpensive, reliable, and easy-to-use contraceptives. Not to be forgotten is the complex system of powerful technologies which have reduced the importance of physical strength in many occupations.

Moreover, busy career women received an "assist" in the form of **prepackaged food** items requiring only a minimum of additional preparation before being served. This new product line was, of course, heavily dependent upon the availability of refrigeration. Also contributing were chains of "fast-food" restaurants, which popped up like mushrooms after a warm rain following the Second World War. By 1990 the decline in "from-scratch" meal preparation, together with the almost complete absence of basic food processing (shelling, shucking, butchering, etc.) at the homesite, seemed to herald the end of what had been a fundamental feature of human life since its very beginnings: economically productive activities at the base camp or dwelling (discussed in Chapters III and IV). But new "high-tech" developments that would reverse this trend were already underway: garage-based "start-ups" and telecommuting.

D. Further Arguments and Issues

The complexity and variability of the relations between technology and society are vividly illustrated by the developments sketched above. Even so, a more detailed examination would reveal additional complications and connections. Indeed, specialists in the areas touched upon would be quick to point out that the above accounts are oversimplified. In any case, some features of the episodes discussed stand out. One of them is a fundamental characteristic of major technological innovations: their emergence in an extended series of incremental steps, illustrated in this chapter by the saga of the "fire drug" in China and in earlier chapters by the development of agriculture, writing and counting, motive power, iron working, and other innovations. Until fairly recently, such series of events took place over a protracted period of time. As long as such changes happened so infrequently or gradually that the majority of the population was either unaffected or unaware of the changes, and as long as accommodations to them were incremental (as they are likely to be under such circumstances), social upheaval was unlikely to result. However, once technological innovation began to take place so rapidly that the lives of large numbers of people were disrupted—as was the case with the traditional textile workers of England—social unrest was much more likely to occur, sometimes accompanied by violence. The difference between these two types of situations is forcefully illustrated by the contrast between the seismic changes that followed the appearance of gunpowder in Europe and what transpired in China where the "fire drug" was invented, a process that took several centuries and was accompanied by incremental changes in areas affected. This case study thus draws attention to a second significant aspect of technological innovation: the rate at which it takes place. It is especially important for matters relating to society.

The accelerating pace of technological innovation and accompanying social change received a lot of attention in the decades following the Second World War. The era's sense of unease and confusion was characterized by Alvin Toffler as "future shock," the result of changes coming so fast that people are kept off-balance psychologically and intellectually. This catchphrase entered the public lexicon by way of Toffler's 1970 best-selling book by that title. Remarkably, this discussion took place before the dizzying pace of innovation in the computer and communication sector had set in! Indeed, by the late twentieth century, these concerns had only grown. In the 1990s Neil Postman articulated his technological determinist concern that Western societies were becoming "technopolies" in which culture is subservient and secondary to technology. Among the factors that have contributed to the accelerating pace of technological innovation are the snowball effect (discussed in Chapters I, II, and V), a growing awareness of the possibility of making and benefiting from devising new ways of doing things, the existence of societies that encourage and support inventiveness and entrepreneurship (discussed above and in the preceding chapter), and the emergence of the powerful partnership of technology and science. One can only speculate how long this accelerating process will or can continue.

1. Technological Determinism, Social Contexts, and Later Developments in Japan

Further insights to be gleaned from the examples discussed above concern the limits of both social contextualism and technological determinism and the ways technology and social and political factors interact with and affect each other. The limits of social contextualism are clearly demonstrated by the example of the rapid adoption of firearms in Japan following their introduction into this Far Eastern country by Europeans, who had a very different culture. Similarly, the limits of technological determinism are demonstrated by Japan's reversion to edged weapons under the Tokugawa shogunate. This development shows that the availability of a powerful technology is not sufficient to guarantee its utilization in any society whatever.

At this juncture it is useful to consider how one might answer the likely response of defenders of technological determinism to this point: namely, that the threat to Japan posed by Western forces with their powerful modern weaponry forced the Japanese to avail themselves of firearms and other modern military technology. Certainly the Japanese sense of vulnerability following the arrival of Commodore Matthew Perry's (1794–1858) squadron in Tokyo Bay in 1853 was well founded. And surely it is correct that technology played an indispensable role in the complex web of developments that culminated in the events of 1853 and 1854 which ended Japan's isolation and marked the beginning of its rapid transformation into a major modern industrial and military power. Without fearsome weapons and an impressive number of naval vessels, Perry could not have sailed into Tokyo Bay and demanded that assistance be given to shipwrecked American sailors and that American ships traveling to the East be allowed to obtain fuel (coal), provisions, and water and thus to call at a few Japanese ports.

But technology *per se* was but one of the factors that contributed to the so-called "opening of Japan." After all, the Chinese invented gunpowder and the basic devices crucial to marine navigation, and they mounted naval expeditions to distant lands during the early Ming dynasty. But they withdrew thereafter into isolation and stagnation as the result of internal social and political factors. Thus, despite China's proximity to Japan and its possession of the relevant technologies, at no time after the Mongols were driven out was Japan threatened by an invasive force from China. Clearly, the mere existence of firearms and marine navigation capabilities did not entail that the Japanese would be confronted on their own shores by powerful foreign forces.

Conditions in the young United States at the mid-point of the nineteenth century stood in marked contrast to those that had prevailed in China since the late Ming Dynasty. In the US it was a time of innovation and entrepreneurship with growth and expansion in almost every dimension. Perhaps most dramatic was the increase in the territorial expanse of the country, with even more area added in the ten-year period ending in 1853 than resulted from the acquisition of the Louisiana Territory in 1803. Immigrants were streaming in and commerce was growing rapidly, aided by steamboats plying the great rivers and constantly-extending railroads. Maritime trade was also increasing in both volume and reach. Perhaps most importantly, the populace exhibited a self-confident mood, a belief in the superiority of the country's institutions and values and in its future growth and prosperity. Furthermore, there was a willingness to take up arms in pursuit of the country's perceived "manifest destiny." This mind-set was very much on display in events leading up to the Mexican War (1848) and in the "Fifty-four forty or fight" slogan of the "hawks" who favored the annexation of the whole of Oregon Country. (In the end, Britain and the US agreed to divide the territory, with the northern half later becoming the Canadian province of British Columbia.)

Evidently, the difference between the threat posed to Japan by China around 1650 (by which time few firearms were to be found in Japan) and that posed by the US in 1853 had little to do with technology as such and almost everything to do with cultural, social, and political factors. The response of the defenders of technological determinism to this point is likely to be that once modern weaponry and marine transport were available, foreign interests were bound to show up (sooner or later) on the shores of Japan seeking treasure, trade, influence, and/or conquest. Consequently, the argument goes, the Japanese really were left with little choice but to modernize and industrialize. That is true, as subsequent events demonstrate, but this too is as much a comment on human societies and people's desire for wealth, status, and power as it is on technology.

In the same decade during which Perry visited Japan, developments were afoot on the adjacent Asian mainland that would culminate in a confrontation between Japan and the expanding Russian Empire. Taking advantage of China's weakness, Russia took possession of a large chunk of Chinese territory at its northeast corner, depriving China of any coastline on the Sea of Japan and giving Russia access to the Pacific Ocean far enough south to avoid a months-long winter freeze-up. The Russians founded the great seaport of Vladivostok at the southern tip of this area in 1860. By no means satisfied, Russia joined other European colonial powers late in the nineteenth century in extracting further concessions from China: two ports on the Yellow Sea at the tip of the Liaodong Peninsula and the right to build and operate a railroad across Manchuria to connect with the Trans-Siberian Railway, effectively bringing Manchuria into the Russian sphere of influence. Western powers were alarmed at the prospect of further Russian expansion in the Far East, as was Japan which had gained control of Taiwan and pushed China out of Korea following a brief conflict (1894–95). Diplomatic efforts to resolve the competing interests of Russia and Japan over Manchuria and Korea failed, primarily because of Russian intransigence. The Japanese withdrew from negotiations, severed diplomatic relations, and (in February 1904) launched a surprise attack on the Russians in the Yellow Sea area. Although few would have expected it, Japan was able to follow up with a series of hard-fought victories over the next year and a half. In the peace negotiated in 1905 by US President Theodore Roosevelt, Russia abandoned its claims in Korea and Manchuria. And Japan, which took possession of Russia's assets in Manchuria, emerged as a major world power firmly launched on its own road to empire.

As improbable as it might have seemed, Japan had succeeded in transforming itself—in a mere fifty years—from a feudal, hermit kingdom into a twentieth-century powerhouse. How was this possible? First of all, the Japanese had to be brought to recognize just how vulnerable they were—no small feat given their isolation. This was accomplished by the events of 1853 and 1854, even though the threat to Japan posed by the United States at the time was actually fairly limited. Furthermore, since it was evident that industrialization and modernization were essential to meeting the challenge facing Japan, and since this could be accomplished most expeditiously by borrowing from the Western nations, it was necessary for the Japanese to reject (at least as far as technology and science were concerned) the earlier dogma (discussed above) that anything foreign was to be shunned because it was a threat to Japanese culture and to the stability of the nation.

Actually, this older tenet was more of a thin veneer serving political purposes specific to an earlier time than a core element of Japanese culture. With the nation stabilized and unified, there was little reason to retain it. Furthermore, an overall characterization of the Japanese would include that they have been quick to avail themselves of the advantages of other peoples' ideas and practices, most notably in the earlier years those of the Chinese. The readiness of the Japanese to adopt foreign inventions is also demonstrated by the speed with which firearms were produced in large numbers in Japan once introduced there. Even Tokugawa Ieyasu, the first Tokugawa shogun, promoted international trade and the development of maritime capabilities. He went so far as to make Will Adams (1564–1620), a knowledgeable English mariner, his trusted advisor on shipbuilding and related matters, as noted above. One must also add the Japanese resentment of the conditions imposed upon them by the treaty of 1854, which made it easier to achieve public support not only of the needed changes but also of the drive to raise the status of Japan among other nations and empires.

The transformation that took place in Japan also illustrates powerfully another important aspect of the technological-innovation-and-society picture: not the effects of technological development on society, but the "social shaping of technology" (that is, the effects of social, political, and cultural factors on it). In 1860, as a first step in coming to terms with the outside world, the Japanese sent a delegation on a globe-circling trip to gather information about the institutions, practices, and resources of other countries. Landing first in San Francisco, the Japanese emissaries subsequently were transported across the Isthmus of Panama by steam locomotive, with which they were very impressed. They then went calling at Washington, London, Paris, Rome, and St. Petersburg. In addition to identifying what their country needed to know about and the sources from which these could be learned, the delegation was charged with recruiting European and American experts to come to Japan to provide instruction and guidance there. In 1867 the Tokugawa shogunate was replaced by a new form of government and the samurai were eased out their dominant position. The new government quickly set about establishing an encompassing educational system, promoting industrialization and economic development, and building a modern army and navy. It should be noted that the Japanese did not merely copy Western ways but learned about them and selected those that served their purposes. One could hardly conceive of a better example of actions taken in the political sphere (indeed, a series of actions) bringing about fundamental changes in the technological repertoire of a people.

2. Invention, New Uses, and Intellectual Property

Paralleling the technology-and-society duality is a related one between the geometrical, physical, chemical, biological, etc., characteristics of a device or process and the use(s) to which it is put. The former factors do not determine for which, when, or under what circumstances a device or process is or will be used, but they do limit its uses, especially with regard to effectiveness and efficiency. At the same time, the function of any technology is part of its identity, as implied by the definition worked out in Chapter II. However, simply using an existing device for a new purpose need not qualify as launching a new technology. For example, using a microwave oven to disinfect a kitchen cleaning sponge rather than to cook or heat food is a new use of an existing technology, not the introduction of a new one. On the other hand, if a device or process is modified in such a way that it is better suited for one of the purposes it serves (but not for another), then one can say that the process of creating a new, distinct technology has begun.

The difference between inventing a new device or process and inaugurating a new use for an existing one is important because it is highly relevant to the question of what is or should be patentable and what is not. As discussed in Chapter VII, patent laws were introduced to encourage people to invest the time, energy, and resources necessary to inventing new devices, processes, and products. The assumption was that the public welfare would be served if innovation were promoted by granting to inventors a monopoly for a limited period of time. Unfortunately, as was also discussed in Chapter VII, the patent system lends itself to exploitation by people seeking to enrich themselves by dubious claims of invention.

The evolution and uses of explosives provide an instructive example of a technology with multiple uses from which sprang two separate technologies. It was suggested in Section B that even though the Chinese used the "fire drug" for two quite distinct purposes, pyrotechnics and weaponry, it nevertheless retained a single identity. That is to say, it was the same technology put to different uses, both of which took advantage of the same special properties of the fire drug. Furthermore, it was seen that both uses were made of it during the period that it was being developed and refined so that technical advances in one area were typically applied in the other, especially during the earlier stages of the process. While there were developments in ensuing centuries specific to both the military and pyrotechnic uses of the fire drug, the basic technology underlying both types of enterprises remained the same. The capabilities that were exploited in skyrockets were the same as those that made military missiles possible. Things changed very little until the middle of the nineteenth century when "smokeless powder" and dynamite entered the picture. Enormously more potent—and dangerous—than the older "black powder," these "high explosives" were the product of that powerful partnership, technology and science, discussed in Chapter II. While black powder continues to be used in pyrotechnics, the new family of explosives supplanted it in warfare, mining, civil engineering, etc.

The use of the same device for different purposes can lead to the development of new appliances. For instance, technicians in radar operations working with magnetrons during World War II found that they could be used to heat water to make tea. Soon thereafter, an American engineer testing a magnetron found that radiation from it would melt chocolate candy, pop popcorn, and cook eggs. From this discovery came the microwave oven, billed as the "radar range." It might seem that this is just another example of one technology being used for two different purposes, but there are good reasons to conclude that microwave ovens and radar tracking devices really are two distinct technologies: (i) they do not exploit the same features of magnetrons and microwaves (one device depends upon the heating effects of microwaves, the other upon their being reflected by various kinds of objects); and (ii) while the development of radar tracking devices was well advanced before efforts to produce a microwave oven even began, a whole new set of challenges had to be met before a marketable appliance was achieved.

The preceding discussion is not intended as a definitive treatment of the issues involved but rather as a starting point for further reflection and exploration. Even the question of what counts as a "different use" of a device is more complex than it might seem initially. Whether or not a particular way of employing a piece of equipment is actually distinct from others depends upon the perspective taken, as can be seen by considering the case of the typewriter. It became a commercial success only after being put to work as a piece of office equipment, which did not happen until typed (as opposed to hand-written) business letters gained acceptance in the business world. Since it was initially marketed as a product for the home (for the use of authors, clergy, serious correspondents, etc.), the typewriter is often cited as demonstrating that the purpose of a device envisioned by its inventor(s) is not always the function that made it economically viable. From a sociological and economic viewpoint, utilizing typewriters in business offices is certainly a different use from that contemplated initially. But these are not different uses from a narrowly technical perspective, which would view them both simply as printing letters and numerals on a piece of paper.

The case of electronic computers is similar but, if anything, more dramatic. These machines were developed by mathematicians, scientists, and engineers for the express purpose of solving very difficult, complex problems in their fields. Given that early computer systems were extremely large and very expensive, and given the presumption that they would be used for tasks similar to those which had led to their development, it is hardly surprising that those who pioneered digital, electronic computing believed that the number of these installations that would be needed was very small. As it turns out, this argument fails on two fronts. Firstly, advances in electronics—the invention of the **transistor**, **printed circuits**, **integrated circuits**, and so on—greatly reduced both the size and the costs of computers. Secondly, as computers became smaller and less expensive, new uses for them were constantly devised—to the point that they have become ubiquitous in the most literal and strongest sense of the term. In contrast to the "number crunching" performed by early computers, more recent ones are used, for example, to monitor the readings of a set of instruments, to determine (on the basis of these) any adjustments needed in the settings of the controls of a device or machine, and to send electronic "commands" to produce those adjustments. This is how braking systems and engines of automobiles currently operate, and it is how the steering of "driverless" cars works. Examples are found in countless other operations, including equipment in the space program.

Because of their complexity and fundamental differences from older technologies, **computer and information technologies** pose special challenges not only in the practical arena but also at the theoretical and conceptual level. Many of these challenges are rooted in what is arguably the most fundamental feature of computing: **the computer program**, "software" as distinct from "hardware." Indeed, the dawning of the computer age can be identified with the implementation of the suggestion (put forward by mathematician John von Neumann [1903–57]) that the instructions specifying the operations to be carried out by a computer be placed in its memory in the form of a sequence of binary numbers rather than being transmitted to it one step at a time as the operations are carried out. Such a sequence of instructions came to be called a "computer program." Creating these programs, especially powerful ones, takes an enormous amount of time and represents a significant investment. Clearly, there would be little incentive to develop elaborate computer programs if there was no compensation for their use by others. Even though it was generally recognized that the public as well as the software creators would benefit from some kind of protection being extended to investments of this kind, achieving this goal was challenging.

In the first place, developing computer programs does not seem to fit into the class of innovations that patent laws were intended to encourage: new devices or machines, new physical or chemical processes, new methods, materials, products, etc. Rather, as some pointed out, computer programs seem to be more like mathematical theorems. Furthermore, copyright laws did not seem to offer the kind of protection needed for investments in software development either. Copyright laws were passed to encourage the publication of books, pamphlets, periodicals, newspapers, etc. in the belief that the public would benefit thereby. Until recently, these laws were understood to protect authors' and publishers' written expression but not the ideas or information contained in documents. Creators of computer programs, however, needed protection not only for specific bodies of coding but also for methods, devised by them, of dealing with problems encountered in program development and ways of extending the range of computer capabilities. Through legislative action and judicial decisions, copyright protection was extended to computer programs, opening the door to the development of the software industry that has played such a major economic role worldwide in recent decades.

This step is but one part of a general trend, during this period, of broadening of legally-sanctioned ownership rights under the heading of "intellectual property." And paralleling what happened in the case of the steam-engine patents (discussed in Chapter VII), this development provided an opportunity for people—in this case an entire class of people—to seek to enrich themselves and/or their clients by filing claims of ownership or infringement. Since there were a lot of questions regarding computer software to be settled—for example, how close to a copyrighted program must another program be in order for a claim of infringement to be sustained—and since so much was at stake in some of these proceedings, lawsuits have become a major part of conducting business in this segment of the economy. Similar motivations lie behind the recent expansion of copyright coverage in areas other than computer software as well as more encompassing protection for patents, trademarks, product brands, etc. The justification put forward for the preponderance of these changes has been that people are entitled to the products of their labor—physical, intellectual, or otherwise. Calls for balancing such claims against the public welfare are muted in comparison, if not drowned out or simply ignored. Furthermore, things have reached the point that a good case can be made that invention and entrepreneurship are being impeded rather than promoted by these "entitlements," heralding, ironically, another area of tension between technology and other elements of human society.

Sources / Further Reading

Andrade, Tonio. *The Gunpowder Age: China, Military Innovation, and the Rise of the West in World History*. Princeton University Press, 2017.

Buchanan, Keith, et al. *China: The Land and the People: The History, the Art, and the Science*. Crown Publishers, 1980.

Busch, Noel F. *The Horizon Concise History of Japan*. American Heritage Publishing Co, 1972.

Cohen, Daniel. *The Last Hundred Years: Household Technology*. M. Evans, 1982.

Cowan, Ruth Schwartz. *More Work for Mother: The Ironies of Household Technology from the Open Hearth to the Microwave*. Basic Books, 1983.

Ebrey, Patricia Buckley. *The Cambridge Illustrated History of China*, 2nd ed. Cambridge University Press, 2010.

Ellul, Jacques. *The Technological Society*. Vintage Books, 1967.

——. *The Technological System*. Continuum Press, 1980.

Enomoto, Tamara. "Giving Up the Gun? Overcoming Myths about Japanese Sword-Hunting and Firearms Control." *The Journal of Research Institute for the History of Global Arms Transfer* vol. 2018, no. 2, 2018, pp. 45–59.

Fairbank, John King, and Merle Goodman. *China: A New History*. 2nd enlarged ed. Belknap Press, 2006.

Fisher, William W., III. "The Growth of Intellectual Property: A History of Ownership Laws in the United States." cyber.law.harvard.edu/property99/history.html.

Hall, Bert S. *Weapons and Warfare in Renaissance Europe*. Johns Hopkins University Press, 2002.

Hane, Mikiso. *Premodern Japan: A Historical Survey*. Westview Press, 1991.

—— and Louis G. Perez. *Modern Japan: A Historical Survey*. Westview Press, 2012.

Hibbert, Christopher. "Ludds, Do Your Duty Well." *Horizon* vol. 16, no. 4, 1974, pp. 58–63.

Hilton, R.H., and P.H. Sawyer. "Technical Determinism: The Stirrup and the Plough." *Past and Present*, vol. 24, 1963, pp. 90–100.

Howell, David L. "The social life of firearms in Tokugawa Japan." *Japanese Studies*, vol. 29, no. 1, 2009, pp. 65–80.

Kelly, Jack. *Gunpowder: A History of the Explosive that Changed History*. Basic Books, 2004.

Kraft, Michael E., and Norman J. Vig, editors. *Technology and Politics*. Duke University Press, 1988.

Levathes, Louise. *When China Ruled the Seas: The Treasure Fleet of the Dragon Throne: 1405–1433*. Simon & Schuster, 1994.

Lifshey, Earl. *The Housewares Story: A History of the American Housewares Industry.* National Housewares Manufacturers Association, 1973.

MacKenzie, Donald, and Judy Wajcman, editors. *The Social Shaping of Technology*, 2nd ed. Open University Press, 1999.

Metropolis, Nicholas et al., editors. *A History of Computing in the Twentieth Century: A Collection of Essays*. Academic Press, 1980.

Mokyr, Joel. *The Lever of Riches: Technological Creativity and Economic Progress.* Oxford University Press, 1990.

Perrin, Noel. *Giving Up the Gun: Japan's Reversion to the Sword, 1543–1879*. David R. Godine, Publishers, 1979.

Postman, Neil. *Technopoly: The Surrender of Culture to Technology*. Vintage, 2011.

Ronan, Colin A. *The Shorter Science and Civilisation in China: 1, An Abridgement of Joseph Needham's Original Text.* Cambridge University Press, 1978.

——. *The Shorter Science and Civilisation in China: 2, An Abridgement of Joseph Needham's Original Text.* Cambridge University Press, 2000.

——. *The Shorter Science and Civilisation in China: 3, An Abridgement of Joseph Needham's Original Text.* Cambridge University Press, 2003.

——. *The Shorter Science and Civilisation in China: 4, An Abridgement of Joseph Needham's Original Text.* Cambridge University Press, 2003.

——. *The Shorter Science and Civilisation in China: 5, An Abridgement of Joseph Needham's Original Text.* Cambridge University Press, 2006.

Smith, Bradley. *Japan: A History in Art*. Simon and Schuster, 1964.

Staudenmaier, John M. *Technology's Storytellers: Reweaving the Human Fabric.* MIT Press, 1989.

Szczesniak, Boeslaw, editor. *The Opening of Japan: A Diary of Discovery in the Far East, 1853–1856*. University of Oklahoma Press, 1962.

Temple, Robert. *The Genius of China: 3,000 Years of Science, Discovery, and Invention.* Simon and Schuster, 1986.

Toffler, Alvin. *Future Shock*, 2nd ed. Random House, 1970.

Volti, Rudi. *Society and Technological Change*, 7th ed. Worth Publishers, 2013.

White, Lynn, Jr. *Medieval Technology and Social Change.* Oxford University Press, 1966.

Chapter IX: Environmental Consequences of Human Inventiveness

JUST AS THE CENTRAL role played by technology in the overall human drama was slow to be recognized, so has an understanding of the part played by technology in the degradation of the environment that has resulted from human activities. Since readers are surely accustomed to a never-ending barrage of news reports, commentaries, and debates concerning the effects of human beings and their technology on the natural environment, it may come as a surprise that public awareness of and concern for such matters is a quite-recent development. This is not to say, of course, that until recently no one had realized that his or her physical surroundings had deteriorated on a sufficient scale to signal a pervasive development rather than simply a local phenomenon. Quite to the contrary. The idea that deforestation and soil erosion had reduced Plato's native Attica to a "skeleton" of its former self is expressed in his *Critias* (111a–112a).

But it was only in the eighteenth century that the basic outlines of the impact that human beings were having on the natural environment began to emerge. This came about as the result, first of all, of systematic studies carried out by the naturalist Georges-Louis Leclerc, comte de Buffon (1707–88). Buffon devoted himself to examining and describing in excruciating—but captivating—detail every aspect of constituents of the natural order. The focus of his work, however, was not on individual entities but on developing a complete description of the kinds of things which they exemplify—in the case of living things, species. For such creatures, a complete characterization needs to include habitat, water and nutritional requirements, pattern of growth, mode of reproduction, etc. Evidently, such accounts have a temporal dimension, hence the name "natural history" or, in French, as in the title of Buffon's great forty-four-volume work, *Histoire Naturelle*.

The passage of time is also central to Buffon's approach to geology, mineralogy, and geography (as these disciplines are currently understood). For example, he did not believe, as essentially everyone once did, that the mountains that dot the surface of the Earth have always existed, more or less the same. Buffon applied his systematic and (as far as possible) exhaustive method to the study of the land: the mountains, hills, plains, streams, rivers, lakes, marshes, forests, grasslands, deserts, etc. His investigations revealed significant differences between uninhabited or recently settled lands and those occupied since ancient times. The latter boasted less forest land, more bare mountains, fewer lakes and swampy areas, and soils of reduced fertility. Buffon's work marks the beginning of Western scientific studies of the impact of human activities on the natural environment.

For some time thereafter, investigations focused on particular examples of such effects, e.g., the consequences of deforestation in the French and Austrian Alps: erosion, floods, siltation, alterations in the courses of streams, reductions in the water level of lakes, and other phenomena. Nothing like an overall assessment of the effects of human activities on the natural world appeared until 1864 when George Perkins Marsh's *Man and Nature* was published. In the late nineteenth and early twentieth centuries, a number of important studies on the topic were made and issues pertaining thereto debated among a few geographers and related specialists. But these had very little influence outside these groups.

During this period, however, there was a growing public recognition of the desirability of preserving certain especially spectacular areas in their natural state, as evidenced by the creation of the first national parks: Yellowstone in the United States (1872) and Banff in Canada (1885). In the 1890s four more national parks featuring towering mountain peaks, waterfalls, and giant trees were created in the Pacific Coast region of the United States: Sequoia, Yosemite, General Grant (later Kings Canyon), and Mount Rainier. The conservation and preservation movements also began to pick up steam at this time. The Sierra Club, for example, was founded in 1892 and soon became embroiled in a hotly debated issue that focused attention at the national level on the conflict between preserving the natural environment and developing it to serve human ends. Around the turn of the twentieth century, the rapidly growing city of San Francisco needed to make provisions for a reliable water supply to meet existing needs and anticipated future ones. The obvious source was the Sierra Nevada watershed 150 miles to the east. The city leaders focused upon a narrow, steep-walled valley through which the Tuolumie River flowed. There they proposed to create a reservoir by building a dam. Since the valley, known as "Hetch Hetchy," lay within the boundaries of the newly-established Yosemite National Park, the city had to seek permission from the federal government to build the dam.

A bitter political battle ensued, separating the preservationists (people committed to saving wilderness areas in their natural state) from the conservationists (those committed to good stewardship of natural resources for the benefit of present and future generations). The former group opposed the project, the latter supported it, as did those who attached little or no importance to its environmental effects. Eventually, the view that seems to have prevailed was that San Francisco's need for water outweighed any losses that might result from building the dam and flooding the valley. In 1913 Congress voted to let the project proceed. Although technology played a crucial role in this drama, it could hardly have been expected to focus attention on the difference that the availability of technologies can make in the impact that human activities have on the natural environment. After all, the Romans built dams to create reservoirs as well as long water conduits ("aqueducts") some 2,000 years ago. The difference in the technologies involved in the two cases would seem to be of little significance.

By the 1960s, it was widely believed that there was an "**environmental crisis**," and the question of its cause(s) received a lot of attention. One viewpoint, shared by a number of those in Europe and North America who addressed this issue, was that people once had lived—and some still did—in harmony with and as an integral part of nature. These thinkers, accordingly, sought to identify the "roots" of this crisis. Given the severity of the environmental degradation that followed in the wake of industrialization and the formation of the powerful science-and-technology partnership, it can hardly be surprising that more traditional ways of living would have seemed absolutely benign in comparison. Before long, however, it became clear to all who studied the matter thoroughly that modern ways of doing things are by no means the only type of human activity to have had a significant impact upon the environment. Indeed, as investigations have proceeded, it has become increasingly obvious that states of affairs once thought to be examples of human beings living in harmony with pristine nature are actually cases of people living in environments they had altered radically. Thus, the first major section of this chapter explores evidence concerning how even early humans may have radically altered their environments and caused extinctions.

A more recent and well documented case concerning the negative impact of powerful new technologies on the natural order is what happened to the American Bison in the nineteenth century. Some 50 million of these huge beasts are believed to have ranged over central North America before the Europeans arrived with their horses, guns, and steel. Early explorers and trappers left reports of herds that stretched to the horizon. In 1850, despite a shrunken habitat and increased predation, an estimated 20 million bison roamed the Great Plains region just east of the Rocky Mountains from Canada's Great Slave Lake to the Rio Grande. But a dramatic change was in the works. In an incredibly short period of time, the vast, thundering herds of buffalo, which once forced steam locomotives to wait while they crossed the tracks, were completely gone! In 1889 only about 500 buffalo could be located anywhere in the US. How could such a thing have happened? Although new technologies played a crucial role, following the pattern discussed above (especially in Chapters VII and VIII), their effects were conditioned by a number of factors: demographic, economic, political, military, and socio-psychological. Because it illustrates this point so well and because the near-extinction of such an iconic species raised public awareness of the gravity of the threat of anthropogenic (human-caused) extinctions, the example of the buffalo is ideally suited for a case study, which occupies section B of this chapter.

This case study is followed by a discussion of four examples of environmental problems that prompted constructive action. The final section of the chapter is devoted to the mixed legacy of the powerful science-and-technology partnership: both wonders and unprecedented environmental degradation.

A. North American Mega-fauna Extinctions in the Late Pleistocene

Prior to the mid-1960s, the prevailing assumption was that people with meager technological resources were too few in number and too limited in weaponry to have contributed significantly to the **extinction of mega-fauna** (large animals). Consequently, the hypothesis that human beings were responsible for the abrupt extinction of most (perhaps 70 percent) of the large (adult weight of 100 lbs. or more) mammalian species of North America some 11,000 years ago was met with skepticism and with suggestions for alternative solutions to this long-standing puzzle. (Eleven thousand years ago is the date long used in most discussions of this phenomenon, but recent radiocarbon methods fix this date at about 12,700 years ago.) Keep in mind, however, that extinctions are a normal part of the on-going life process. As new species arise and compete with existing ones for ecological niches, extinctions are inevitable. But that was not what happened in North America in the **Late Pleistocene Epoch**, a time when the continental ice sheet was retreating northward. Rather, the places in the ecological web once occupied by the extinct species were, by and large, simply left vacant.

A number of possible explanations for the extinction of the mega-fauna in North America were put forward, including the impact of an extraterrestrial body (an asteroid or comet), a deadly pandemic, climatic changes, and the activities of human beings. Nothing much ever came of the first two of these hypotheses, but the fourth one sparked an extended debate. The basis for the suggestion that human beings were responsible for the extinctions was (i) their occurrence relatively soon after the arrival of people in North America and (ii) the existence of clear evidence that these newcomers hunted the animals. Those who thought that this explanation lacked credibility tended to favor the view that climatic changes were the primary cause of the extinctions. They emphasized that many rapid changes in climate took place during the Late Pleistocene, which would have put animal populations under stress, and suggested that human beings may have done little more than finish off the remnants of species already under threat of annihilation. They also pointed out that human beings had coexisted with Africa's great variety of large animals for many thousands of years without there having been a wave of extinctions like that which occurred in North America.

Admittedly, it might seem that the existence of so many species of large animals in Africa, despite the presence of human beings for many thousands of years, would cast doubt on the suggestion that people were responsible for the abrupt extinction of such a high percentage of the mega-fauna in North America. But the matter is a bit more complicated than it might seem. First of all, there was a limited but significant number of extinctions among large-animal species in Africa during the (pre-modern) period in which human beings or other members of the ***Homo* genus** were present, with some indications of a transitory spike at the time when a new and better way of making stone tools made its appearance: the flake technology (discussed in Chapter III). Secondly, there are good reasons to expect extinction rates among mega-fauna to be less elevated even when human beings are present if certain conditions are met: specifically, if the animals have co-existed with people (or other members of the *Homo* genus) long enough to have learned or been programmed by natural selection to fear them and/or to follow other survival strategies. In Africa there was enough time for such defensive strategies to develop before members of the *Homo* genus were sufficiently numerous and possessed of sufficiently effective hunting weapons and techniques to make an impact—but not in North America in the Late Pleistocene.

Of course, the wave of extinctions that swept North America would not be expected to have taken place immediately after the arrival of human beings but only after their numbers had grown and their hunting capabilities honed. Indeed, there is a growing body of evidence that the earliest human inhabitants of the Americas, who are believed to have arrived about 15,500 years ago, were not big-game hunters but hunter-gatherers who subsisted on such things as roots, nuts, crustaceans, fish, birds, small reptiles, and small mammals. Even so, that picture clearly had changed by about 13,000 years ago, a time when a very distinctive type of stone spear point, called a "**Clovis point**," appears to have been used in the pursuit of large animals. The first of these to be found (near Clovis, New Mexico) were embedded in a cache of bones belonging to a sub-species of extinct giant bison. Typically around four inches long and one third of an inch thick, these (often) superbly fashioned points feature a shallow groove extending from the squared-off butt-end toward the tip, a characteristic that would have facilitated mounting the points on wooden shafts. Some 10,000 of these points have been found in locations scattered across most of North America. So distinctive are they that they must have been produced in essentially the same way, a technique that seems to have appeared quite suddenly and rapidly spread across the continent. Examinations of collections of Siberian artifacts have revealed no hint of a tool-making tradition of this kind. Thus, it seems reasonable to conclude that the Clovis point is an American invention. As remarkable as the sudden appearance and rapid spread of this new technology is, its abrupt disappearance some 400 or 500 years later, leaving only enticing artifacts, is even more so. At this point, one can only speculate as to what happened. In any case, this new technology certainly enhanced the hunting capabilities of the early Americans; and the close match between the dates of the Clovis phenomenon and that of the mega-fauna extinctions lends support to the argument for a human role in these extinctions.

If wholesale extinctions had occurred only in North America, and if small animals and plants had been similarly affected, the matter would have been more of a local puzzle than something of global significance. But such was not the case. Small animal and plant species generally did not become extinct at an abnormally high rate at those places and times when the numbers of mega-fauna species were in steep decline. Furthermore, reductions in mega-fauna species also took place in South America (roughly 1,000 years later than in North America), in Australia (much earlier, about 15,000 years ago), in New Zealand (much more recently, 900 years ago), and Madagascar (even more recently, about 800 years ago). In every case, these large-scale extinctions occurred after the arrival of human beings, in some instances within a relatively short period of time, in others only after some time had elapsed, depending upon certain factors. The very nature of the habitats of some species of animals, small as well as large, renders them especially vulnerable to rapid extinction when human beings arrive upon the scene. As discussed above, isolated islands are a prime example of such states of affairs. Under some conditions, on the other hand, mega-fauna species may escape extinction, or the threat of it, because all or some of their range is inhospitable to human beings. But the viability of a region as a place for human habitation can change as a result of technological development, as is illustrated by the invention of the sewing needle together with the tanning of animal hides, which enabled people to establish themselves in central and northern Eurasia. Similarly, as discussed below, the introduction of the domesticated horse into the Americas made possible the "buffalo economy," which in turn supported the rich Native American cultures of the Great Plains. Furthermore, the vulnerability of a species to extinction and the timing of it (if it happens) can be affected by climatic changes, as critics of the Pleistocene Overkill hypothesis have pointed out.

As more and more factors are identified that might affect the significance of the presence of human beings at times when meta-fauna extinctions took place, the more difficult it becomes to make headway regarding the matter. Especially problematic is the lack of precise information regarding the timings of extinctions as well as the size of human populations at a given time and place in the distant past. Rather than focusing upon the timings and possible causes of individual extinctions (or groups of them), a promising new approach involves making a country-by-country, worldwide comparison of the correlations between late Quaternary mega-fauna extinctions and (i) climatic changes and (ii) the arrival of archaic or modern human beings in a particular geographic location. During the period from 132,000 years ago to 1,000 years ago, the study found, 30 percent or more of large mammalian species became extinct (globally or continentally) in those places (North America, South America, and Australia) where the animals had not encountered members of the *Homo* genus until the relatively recent arrival of modern human beings. On the other hand, in Africa (where the *Homo* genus originated) and in southern Asia (which was populated by *Homo erectus* beginning as early as 1.8 million years ago), extinction rates among large mammals were relatively low during this same period. This difference is to be expected, as explained above. The one major exception to this pattern is southern and western Europe, which suffered losses of mega-fauna species during the timespan of the study comparable to those of the Americas and Australia even though *Homo erectus* had arrived there at about the same time as in southern Asia. During this period, Europe and the rest of the planet underwent extreme climatic changes. The last interglacial (the warm interval between the most recent period of glacial formation, "ice age," and the one preceding it) began about 132,000 years ago and lasted some 40,000 years. The cold period that followed it ended about 10,000 years ago. Changes in temperatures and precipitation as well as the rapidity with which these changes took place were analyzed in this study, which concluded that (overall) the severity of meta-fauna extinctions between 132,000 and 1,000 years ago are closely linked to the presence of modern (or archaic) human beings, and that only in the case of Eurasia is there even a weak connection between the extinctions and climatic change. Although these findings are sure to be challenged and many questions remain (for example, what accounts for the departure of southern and western Europe from the dominant pattern?), and even though the effects of human activities are heavily conditioned, it is becoming increasingly clear that the inventive and self-transforming human being is a disruptive presence on the face of the planet. Few cases make this point so clearly as the near-extinction of the buffalo in North America in the late nineteenth century.

B. Case Study: The Vanished Thundering Herd

The technological factors that played a more-or-less direct role in the complex drama that culminated in the decimation of the great herds of bison include **a domesticated animal (the horse)**, the **breech-loading rifle**, the **steam locomotive**, and **refrigeration**. Horses were introduced into North America by the Spanish. At some point, a wild population of the animals became established on the Great Plains, likely originating with horses that had escaped from or been abandoned by the Spanish. The Native Americans managed to capture and train some of these, and they learned to breed horses as well. The result was a transformation of Native American life on the Great Plains. To the hunter, buffalos represent a tempting supply of meat; but preying upon them on foot, armed only with stone-tipped arrows and spears, is not a very promising undertaking. One ingenuous stratagem employed by the Native Americans to get within bow-and-arrow range of the animals was to cover themselves with wolf pelts and to approach on hands-and-knees. Another effective technique was to stampede herds of the buffalo over cliffs or into ravines, killing them, crippling them, or otherwise rendering them more vulnerable. Opportunities to exploit this strategy, however, were limited, as were other sources of food, with the result that life on the Great Plains tended to be a matter of "feast or famine." Consequently, the region remained sparsely populated prior to the arrival of horses. However, with the advantage of well-trained mounts (and iron arrowheads and spear points, when they became available, and eventually guns), the Native American hunters were able to avail themselves of the food supply that the great herds of bison constituted. Bison became their primary source of nourishment, and they fashioned tents, robes, and other articles from their tanned hides and made good use of other parts of the carcasses. They even used dried buffalo droppings for fuel. So successful was the new "buffalo economy" that large numbers of the Native inhabitants of adjoining regions abandoned their traditional ways of living, which included growing crops, and took up the pursuit of the buffalo out on the prairie.

As seen in Chapter V, the domestication of animals and the development of ways of making use of them are an important part of the human technological repertoire. Of course the "wild horses" captured by the Native Americans were not "wild" in the same sense that zebras (which have never been domesticated) are wild. They were descendants of Eurasian horses that were domesticated some 5,000 years ago and bred selectively thereafter. And they had not been in a feral state long enough for natural selection to have resulted in significant changes in their genetic makeup. What the Native Americans did was to build on this foundation, developing the skills and practices that made them such superb riders.

A second technological contribution to the destruction of the buffalo herds was the invention of the **breech-loading rifle** with its pre-packaged ammunition in the form of metal-cased cartridges. These made it possible to reload and fire while riding a horse, a feat hardly imaginable with earlier muzzle-loading firearms. A lighter, shorter rifle, called a "**carbine,**" was produced specifically for use on horseback (see Figure 9-1 below). Cavalry units were armed with it during the American Civil War, but it was not generally available until after that conflict. The advantage of such a weapon for the buffalo hunter hardly needs explaining.

Yet another technological development that played an important role in the demise of the bison herds was the extension of the **railroads** west of the Mississippi

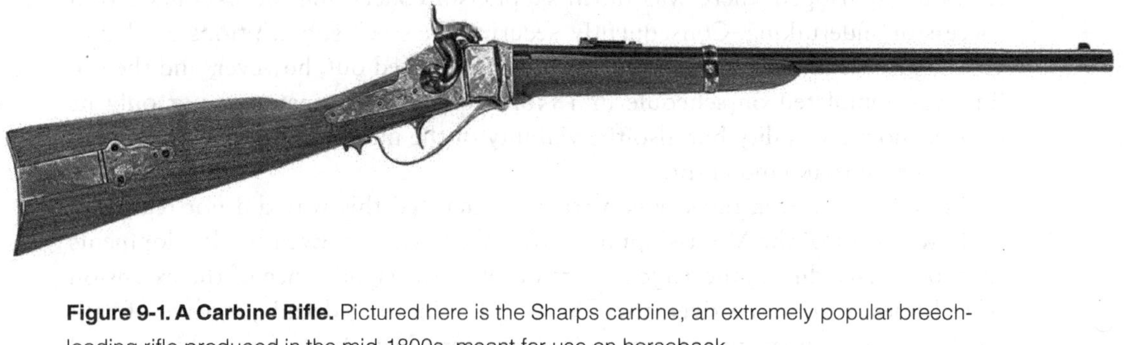

Figure 9-1. A Carbine Rifle. Pictured here is the Sharps carbine, an extremely popular breech-loading rifle produced in the mid-1800s, meant for use on horseback.

River following the Civil War. Prior to that, the construction of bridges across the Mississippi and Missouri rivers had been held up by engineering challenges, the cost, opposition from the steamboat operators and their allies, and overall poor prospects for a profitable enterprise. Most importantly, rail lines crossing the river into the sparsely populated region to the west simply could not be assured of generating enough revenue to justify building them. And without access to markets, few people were going to move into the region to carve out farms and ranches or to launch commercial, extraction, or manufacturing operations. The drive to overcome these obstacles drew support stemming from a variety of motivations: the desire to make a profit, the hope of having a better life by making a fresh start, the attraction of being a part of the fulfillment of the "manifest destiny" (discussed in Chapter VIII), the "itch" for adventure, the allure of a romanticized vision of "the West," and so on. Believing that settlers would move into wilderness areas if transportation and access to markets were provided, government leaders and businesspeople worked out an arrangement whereby railway companies would be granted not only rights of way for the tracks but also blocks of land along the routes which they could sell, once the railroads were built, to recover the costs of their construction—and to turn a profit.

However, the first railroad land grant, authorized by the US Congress in 1850, was not for an east/west route but for a north/south one running through central Illinois. The last Native American group having been removed to the west by 1833, large numbers of settlers were flocking into the state and were soon demanding that something be done about the primitive state of the roads. The question, of course, given the relatively undeveloped condition of Illinois at the time, was how could the needed improvements be financed? The solution that eventually emerged was the formation of a railroad company which contracted to build a railway the length of the state on terms like those described above. But even with swarms of potential customers pouring into the state and with no great rivers to be bridged, there was much skepticism concerning the likelihood of a successful undertaking. Consequently, securing the stock subscriptions needed to get the project underway was not easy. Things worked out, however, and the rail line was completed on schedule in 1856. It quickly demonstrated not only its own economic viability but also the viability of the method employed in launching it: the railroad land grant.

As noted, however, railway construction financed this way did not really get underway west of the Mississippi until after the Civil War. Even so, developments that took place during the 1850s are relevant to the significance of the extension of rail lines west of the river after the war. During that decade, a wave of railroad construction swept the region from western New York, Pennsylvania, and

Maryland to the Mississippi River. Almost overnight, rail lines stretched without interruption from New York City, Philadelphia, and Baltimore to the "Father of Waters," as Abraham Lincoln (1809–65) termed it. The region underwent an explosive growth in population, agricultural production, commerce, and industrial output, as evidenced by the city of Chicago, which quickly became a center of rail traffic without peer. From a population of less than 5,000 in 1840, it grew from about 30,000 in 1850 to some 112,000 in 1860. When the Civil War ended, there were reservoirs of people and resources just east of the Mississippi, poised for a surge across it onto the plains.

From 1865 to 1890, railroad construction west of the Mississippi proceeded at a frenetic pace, epitomized by the rate at which track was being laid as the Central Pacific/Union Pacific transcontinental railway neared completion: three miles of track per day, with a record ten miles laid one day on a bet. Of course, surveying parties, bridge builders, and graders had done their work well ahead of the track layers. The transcontinental railway, which was financed by the land-grant method, was completed in 1869. In Canada the first rail connection to the Pacific coast was completed in 1885. The extension of the rail lines westward was accompanied by an influx of adventurers, big-game hunters, prospectors, miners, and merchants as well as farmers and ranchers. As a way of realizing a return on their investments, rail companies went to great lengths to recruit people, sending agents as far as Russia to persuade them to come and establish farms and businesses along the rail lines.

However, the presence of more people needing food and the availability of breech-loading rifles and horses put more pressure on the buffalo, especially since very wasteful practices were followed in utilizing the animals killed. To make matters worse, buffalo robes were very much in demand back in the East, so that it became profitable to hunt the huge beasts just to obtain the hides. As a result, countless buffalo were slaughtered for that reason alone, with little use made of the rest of the carcasses, which were simply left to rot. The cattle farmers were another source of economic—and hence political—pressure on the buffalo herds. They wanted the open rangeland, then in the public domain and available for pasturage free of charge, for their cattle. The extension of the railroads into the region had made it possible to transport cattle over the immense distance that separated the grazing lands of the Great Plains from the markets in the growing urban centers of the northeastern US. Shipping operations of this kind had been underway from the Midwestern farming areas east of the Mississippi prior to the Civil War, but the greater distance that had to be traversed to move livestock from the Great Plains to the eastern markets took its toll on the cattle in the crowded rail cars and on the profits of the cattle farmers.

The invention of the **refrigerated railcar** (1877) led to dramatic changes (see Figure 9-2 below). Previously, it had been possible to ship dressed beef over significant distances only during the winter. Furthermore, the prevailing practice at the time—as it had been in centuries past—was for beef to be purchased at the local butcher shop, where the animals were slaughtered and carved up. All this was to change with the advent of refrigeration, greatly spurred by interest and investments from Gustavus Franklin Swift (1839–1903), founder of the Swift & Co. meat packing company, which gradually transformed the meat packing and shipping industries. At first this system relied upon bins of ice in the insulated rail cars, but this method of keeping beef cold soon was replaced by mechanical systems. Refrigeration made it possible to ship dressed beef over great distances. Thus, beef produced on the Great Plains could be brought to the dinner tables of the growing population centers of the East without the expense of shipping whole animals the entire distance. As a result, it could be sold at a lower price. That, in turn, led to increased sales and higher profits for the cattle farmers. Consequently, the practice of shipping beef "on the hoof" to the East soon came to an end. Chicago, the western terminus of the first refrigerator car operation, quickly became the center of the meat-packing industry, a development that contributed to its becoming the preeminent rail center of the central United States.

As the profitability of raising cattle on the rangelands increased, so did the political pressure to remove the buffalo from the Great Plains. The railroads also

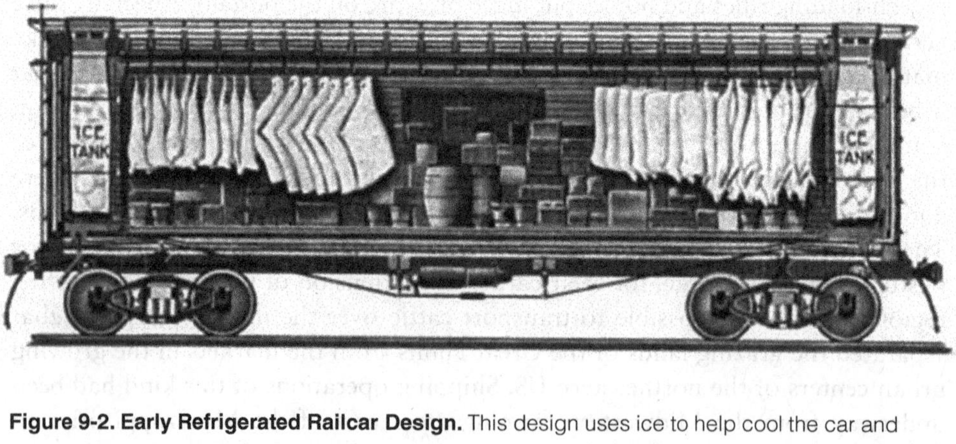

Figure 9-2. Early Refrigerated Railcar Design. This design uses ice to help cool the car and keep its contents from spoiling. Mechanical refrigeration technology (like that found in modern-day refrigerators) would not be widespread in railcars until the mid-twentieth century, though the technology itself had been developed decades earlier.

supported the assault on the bison herds because they held up the trains and damaged equipment. Even so, the buffalo faced an even greater threat deriving from the strategy employed by the US federal government and army in the struggle to subdue the "Plains Indians," i.e., the Native peoples who had adopted the buffalo-hunting way of life described above. Inevitably, the westward movement of prospectors, miners, ranchers, homesteaders, etc. brought them into renewed and repeated conflict with the Native Americans. More often than not, hostilities resulted from infringements upon lands recognized in treaties as reserved for these Native peoples. Time and again, injuries, fears, and retaliations escalated into open warfare along the western frontier from the 1860s to the 1890s. Unfortunately for the buffalo, a core element of the basic strategy followed by the US Army in these "Indian Wars" was to destroy the economic basis of armed opposition to it. In the case of the conflicts with the Native Americans of the Great Plains, this meant the destruction of the bison herds. But the primary role of the army in this process was not actually slaughtering buffalo: it was supporting and defending those who did. Governmental involvement was less of a factor on the Canadian side of the border, in no small part because railroads and the ensuing population growth did not arrive west of the Great Lakes until the herds of plains bison had been decimated already. Even so, the hunters and hide merchants brought about a similar result there.

The destruction of the bison herds in the US was but part of a campaign of what has been called "total war," i.e., military operations directed not only to defeating opposing combat forces but also to destroying utterly the enemy's ability and will to offer resistance. Here, "enemy" is understood to include entire populations: men, women, and children. In such a context, avoiding casualties among non-combatants is not a priority. Rather, they are subjected to assaults and/or deprivations as part of an effort to weaken the opposition in every way possible. Of course, there had been a history of violence between European colonists and Native Americans as soon as the latter arrived in North America (though the colonists were responsible for much of it). And dwellings and resources of all kinds had been destroyed in attacks by both groups. But, in comparison, the campaigns to subdue the Native peoples of the western US during the latter part of the nineteenth century, taken together, were unprecedented, particularly in scope and systematic character. And this was not an accident. Both the architects and the executioners of the "total war" approach to subduing the Native Americans of the West were officers (or former officers) of the US Army, flush with victory in the Civil War. That victory was secured in no small part by the vigorous application of the principles of "total war" as set out in General Order No. 100, April 1863. The authors and chief proponents of this policy were two Union generals, Ulysses S. Grant (1822–85) and William T. Sherman (1820–91).

By the early 1870s, the bison population was plummeting, awareness that extinction was on the horizon was growing, and movements to prevent it were getting underway. Meanwhile, Grant had been elected President; Sherman had succeeded him as Commanding General of the US Army; and General Philip H. Sheridan (1831–88), who had been Sherman's chief lieutenant when he was in command of the plains division, succeeded him in that position. In 1874, the US Congress passed a bill placing the buffalo under federal protection. Sherman and Sheridan urged Grant not to sign the bill on the grounds that the only way that the Plains Indians could be civilized, i.e., persuaded to settle down as peaceful farmers, was to eliminate the buffalo way of life as a possibility. They would then have to accept the terms of the government in Washington or starve. And if some of them did, the attitude seems to have been, so much the better. After Congress had adjourned, Grant allowed the bill to lapse without his signature, a maneuver known as a "pocket veto." When the matter came up again the following year, Sheridan carried the same message to a joint session of Congress and garnered enough support among the "hard liners" to forestall further action. The racist views prevalent at the time surely contributed to this result. Evidently, the complete loss even of such an iconic element of the natural environment as the American bison did not outweigh the benefits seen to accrue from an extreme military strategy. No effective action that benefited the buffalo was taken by the US federal government until the passage in 1894 of the Yellowstone Protection Act, which outlawed hunting, timber cutting, and extracting minerals in the park, established penalties for violations, and made provisions for enforcement—too late to have saved the buffalo.

Fortunately, ranchers and other individual citizens in Canada and the US had already taken steps that would insure the preservation of the bison species. They had established private herds of the animals by rounding up buffalo calves (strays, orphans, etc.) and raising them. As these herds grew, they were traded and divided. Because of their huge size and strength, managing and moving bison is a challenging undertaking. In 1906, the largest buffalo herd in North America, which had been assembled by two Montana ranchers, was offered to the US government, but Congress declined to appropriate funds for the purchase. A deal was worked out with the Canadian government, however, and the daunting task of moving the seven-hundred-strong herd north to Alberta began.

The example of the destruction of the bison herds clearly illustrates just how complex technology-and-the-environment subjects can be. All four of the technological innovations discussed above (domesticated horses, breech-loading rifles, railroads, and refrigeration) were crucial to the confluence of factors that conspired to dethrone the buffalo. Equally important were the many demographic, economic, political, military, and sociological developments involved. Furthermore, the relations among these factors can be rather complicated. For example, absent the introduction of domesticated horses into North America quite some time in the past, there would have been (in the nineteenth century) no large population of Native Americans dependent upon the "buffalo economy" and hence no reason to support the slaughter of the bison herds as a means of forcing those fierce warriors to accept the terms demanded by Washington. Similarly, had it not been so fashionable to wear a buffalo robe or coat, there would have been much less economic motivation to venture onto the Great Plains to kill and skin buffalo and to tan the hides. And these are just two of many such examples.

C. Corrective Action Prompted by Over-Exploitation of Natural Resources

The Congressional approval of a bill to protect the buffalo in the 1870s and the founding of private herds during that period attest to a growing recognition of the harm to the natural order that human activities can cause and a willingness to do something about it. But such awareness and concern were not sufficient to prevent the complete removal of what was once another major feature of the North American scene: the enormous flocks of passenger pigeons that blackened the sky much as the buffalo had darkened the prairie. As late as 1800, the passenger pigeon was the most numerous species of North American bird. Highly sociable, they roosted in such incredible numbers that they broke large tree limbs. In nesting season, they gathered in huge colonies, building dozens of nests in a single large tree. Because they congregated so densely, whether feeding, flying, roosting, or nesting, they were an easy target for hunters. Given this behavior pattern and their enormous numbers, these pigeons were an important source of meat, not only for the Native Americans but also for the European colonists and their descendants, until the precipitous decline of the pigeon population in the latter part of the nineteenth century.

Among the wide variety of methods used to capture these birds were bombarding them with sticks and stones, snaring them in large nets, and knocking them from trees with long poles. (Juveniles not yet able to fly were especially vulnerable to this tactic.) As industrialization and urbanization proceeded during the nineteenth century, more powerful technologies were utilized and hunting and trapping pigeons became increasingly commercialized. Especially notable is the introduction of the breech-loading shotgun. Using the double-barreled version, one could bring down a dozen or two birds with a single discharge of both barrels. Another important technological development was the construction of the railroad network that provided access to markets in major cities. Pigeons were shipped into New York and Chicago by the carload. As early as the 1850s, lawmakers in a number of states attempted to do something about the decline in the passenger pigeon population but got nowhere. A formidable barrier to effective state action was the migratory nature of the species. Efforts to breed the birds in captivity were unsuccessful as well, and the last known member of the species died in a zoo in 1914. Although public concern for the fate of the passenger pigeon failed to prevent its extinction, the attention it drew energized the movement to end the indiscriminate killing of birds. In 1918, following negotiations between Canada and the US, the Federal Migratory Bird Treaty Act went into effect in the US. It and a corresponding Canadian law were major milestones in the campaign to reduce the environmental degradation that all too often is associated with human technological and economic development.

The fate of the passenger pigeon and that of the American bison illustrate cogently what happens all too often when powerful new technologies become available for taking wild game or fish and for transporting the products obtained to distant, large-scale markets: the population of the targeted species collapses, wiping out its value as a natural resource and putting it on the road to possible extinction. A recent, even more disturbing example is afforded by the seismic changes that swept through marine fisheries around the world in the third quarter of the twentieth century. Early in that period an entirely new type of fishing vessel began to appear at those places where commercial fishing was concentrated because of the abundance of fish. These enormous "**factory trawlers**" (300 feet or more in length) were designed for fishing in waters far from their home ports. They towed huge trawl nets behind them which, when full, were hauled up their stern ramps. These leviathans could scoop up as many fish in an hour as older trawlers could hoist aboard in weeks. The catch was immediately processed on board and frozen. Since there was no immediate urgency about getting this frozen cargo to port, these factory ships could remain at a given fishing site for days or weeks—or until they were either fully loaded or ready to move on to another site. The Soviet Union, Germany, Poland, and Spain dispatched these giant trawlers to fisheries around the globe. And from the viewpoint of functionality and efficiency, they were a rousing success—so much so that they quickly put the very foundation of the marine fishing industry in jeopardy.

The problem was that the factory trawlers were doing such a thorough job of catching fish that not enough were being left to reproduce in sufficient numbers to prevent the population from collapsing. Furthermore, since the best marine fishing sites were in international waters, lying outside the jurisdiction of individual countries, and since there was no international agency capable of overseeing and regulating the situation, there was nothing to restrain or even discourage the operators of the factory ships from taking every fish they were able to ensnare. Worse still, there would have been no point in any particular factory ship's limiting its catch at a given site, since any desirable fish left would have been scooped up by another of these huge vessels. This situation exemplifies what has come to be referred to as the "the tragedy of the commons," the destruction of a shared resource as the result of its over-exploitation in a context in which there is no framework for collective action and no reason to think that an individual's exercising restraint will have any effect on the long-term fate of the resource. Of course, it is theoretically possible in such situations for the interested parties to agree to an arrangement that protects the resource through agreed-upon limits to its utilization. In practice, however, such agreements are notoriously difficult to achieve and even more difficult to make work.

Not surprisingly, a resolution of the problem posed by the factory trawlers did not come about in this way. They were reined in—indeed, put out of business—by actions taken by individual nations taking advantage of the locations of the good fisheries of the world: shallow areas of the oceans relatively near landmasses. In response to the protests of established, land-based fishers regarding the factory trawlers' taking so much of the fish, countries with good fisheries near their shores (including the US and Canada), assumed jurisdiction over fishing operations in "exclusive economic zones" extending out two hundred miles from their coastlines. Once restricted by quotas, if not completely excluded from these fishing sites, the factory trawlers were no longer economically viable because it cost so much to operate them. By 1980, they were being taken out of service, bringing to an end another episode of the dark side of human inventiveness: the havoc wreaked on elements of the natural environment—all too often to the detriment of human beings themselves. An arresting aspect of this case is that the factory trawlers contained within themselves the seeds of their own destruction. Had no restrictions been placed on fishing in the exclusive economic zones, the factory ships would have quickly decimated the desirable species at the good fisheries and put themselves out of business. Unlike domesticated horses, breech-loading guns, locomotives, and refrigeration systems, factory trawlers had only one purpose, which they served efficiently only under a very narrow set of conditions.

Unfortunately, the arrival of the factory trawlers at the marine fisheries does not mark the beginning of the degradation of the fisheries from over-exploitation nor does the disappearance of these giant fishing vessels mark the end of it. This point is illustrated powerfully by the fate of the Grand Banks and the Georges Bank, two of the richest fisheries in the world, which lie off the southeast coast of Canada and New England. Earlier technological innovations together with growing numbers of fishing operations were already having a negative impact on these sites before the factory trawlers arrived. During the nineteenth century, fishing vessels were wind-powered and fish were caught on baited hooks and hauled over the side rails by means of gaffs. Beginning around 1900, the marine fishing industry expanded worldwide and became more efficient with the development of more modern equipment. Especially significant was the introduction of diesel engines which made it possible for fishing boats to move about at will and thus to scoop up immense quantities of fish by towing huge trawl nets. These engines also made it possible to employ power-driven equipment to handle the enormous nets that came into use and to hoist the masses of ensnared fish aboard. The use of airplane spotters, radar, and sonar to locate schools of fish was also important, as were weather forecasting and communication by radio-telephone.

Of course, these innovations and the expansion of commercial fishing resulted in a dramatic growth in the total amount of fish brought in, as illustrated by the thousand-fold increase in the tonnage recorded by New England fishers between the middle of the nineteenth century and the middle of the twentieth. But there were clear signs of trouble ahead. One early signal was the downward trend evident in the size of the fish being caught. Another was that, despite increasing fishing capacity and activity, the moving average of the annual catch of the New England fishers began dropping about 1950, significantly before the factory trawlers started to make an impact in the mid-1960s. The underlying problem of more and more people going after a finite natural resource with ever more powerful technologies showed itself clearly when the huge vessels departed: the total catch brought ashore by the New England fishers continued to decline. By the early 1990s, it plunged so steeply that the US government declared a complete moratorium on fishing in part of the Georges Bank in order to give the population of desirable species a chance to rebound. For the same reasons, Canada shut down the Grand Banks fishery even though 40,000 jobs were at stake.

The contrast between the attitudes expressed and the actions taken when extinction threatened the American bison and the passenger pigeon and those evoked by the looming demise of the North Atlantic fisheries testifies to the depth of the changes in the attitudes and priorities that took place during the twentieth century. These changes were fueled in large part by a series of natural disasters, the impact of which was enhanced by best-selling books, iconic photography, and movies. An early, especially notable example is the "dust bowl" of the 1930s, a drought- and wind-stricken area that included the panhandles of Texas and Oklahoma and adjacent parts of Kansas, Colorado, and New Mexico. The most horrendous feature of the dust bowl was the dust storms, monstrous clouds of wind-driven dust in which it was impossible to see or even breathe. Stripped of its native sod cover and loosened by cultivation, the soil was picked up by the fierce winds, resulting not only in dense dust clouds but also in severe soil erosion. In desperation, a third of a million farm workers pulled up stakes and made their way as best they could to California. This saga was portrayed in John Steinbeck's novel, *The Grapes of Wrath* (1939), and John Ford's Academy Award-winning (1940) movie based on it.

The dust bowl disaster prompted the recognition that land, like other natural resources, could not be exploited at will without painful consequences and that land use must be carefully managed if people are to reap the benefits which it is capable of providing. In 1933 the Soil Erosion Service (later the Soil Conservation Service) was established to advise and assist farmers in preventing soil erosion caused by wind and water. The agency set to work promoting such conservation strategies as check dams, crop rotation, contour plowing, and strip cropping. The dust bowl and soil erosion differ significantly from earlier examples of the negative effects of human technological activities and the responses elicited by them. In this case, (i) human life and health were affected directly and (ii) general features of the natural world (soil and air) were involved, not just a single species or type of resource.

The movement of environmental concerns toward center stage was interrupted by the Second World War but resumed with a vengeance when it ended. Indeed, because the war effort took precedence over nearly everything else, it can hardly be surprising that the state of the environment was considerably worse in 1945 than it had been in 1939. The first headline-grabbing postwar environmental disaster exemplified both characteristics noted above. It also exhibited another basic feature that was to be a central focus of environmental awareness and action: the indiscriminate dumping of wastes of all sorts into the air, streams, lakes, and oceans, and onto the land. It had been known for centuries that burning coal, smelting ores, and related activities release dirty and foul-smelling smoke and fumes into the air. And there were reasons to believe that breathing air thus polluted was not only unpleasant but also unhealthy. The severity of the threat posed by air pollution showed itself in starkest terms in late October 1948 in Donora, Pennsylvania (see Figure 9-3 below).

This small industrial city lies in a deep, narrow valley on a horseshoe bend of the Monongahela River. Humidity tends to be high and fogs are frequent, as is typical of river valleys in rugged terrain. Industrialization added coal smoke,

Figure 9-3. Pollution in Donora, Pennsylvania. This photograph shows the high level of pollution emitted by industry in Donora in 1910. This grew worse until a combination of pollution and meteorological factors resulted in the 1948 disaster.

automobile exhaust, and smokestack emissions. In the mid-twentieth century, the US Steel Corporation was operating a zinc works and a steel and wire mill at Donora. (Zinc is used to coat sheet metal, pipe, and other iron and steel products to prevent them from rusting.) Among the pollutants released by these plants were hydrogen fluoride and sulfur dioxide. When these gases are mixed with fog, the smog that results is thick, yellowish, and corrosive. In late October 1948, meteorological conditions in Donora prevented the pollutants from being carried aloft and dispersed by air currents. The city had fallen victim to a thermal inversion, a reversal of the usual pattern of temperatures dropping with increasing altitude. In a temperature inversion, a layer of cold air lies under warm(er) air. The usual thermal process is for air to be warmed through contact with rocks, plants, bodies of water, etc. at the Earth's surface. Thus warmed, it is lighter and rises, while the un-warmed, cooler air above settles lower. The circulation thus set up serves to disperse everything from the smoke of campfires to industrial emissions. When cold air is somehow trapped under warmer air, this circulation does not take place and pollutants released at the lower level are not carried away but remain in the vicinity and can build up to toxic levels.

And that is what happened in Donora. By the second day of "the smog," people were coughing and exhibiting other symptoms of respiratory distress. The situation worsened with each passing day. The healthcare resources of the city came under severe stress. Oxygen to bring relief to the worst cases was brought in from surrounding communities by emergency workers. The smog became so thick that visibility was essentially zero and driving automobiles was all but abandoned. By the time rain brought relief on the fifth day, twenty people had died and from one third to half of the population had been sickened.

The Donora incident is just one example of the disasters that focused attention in the postwar period upon the necessity of bringing air pollution under control. In 1952 an especially thick smog settled on London. Within a period of only five days, 4,000 people had died. In Tokyo during the postwar period the air became so polluted that despite wearing protective masks police were able to remain outside directing traffic only for short durations. The scope and complexity of the problem quickly elbowed its way into the public consciousness and has remained there as new aspects of it have come to light and more countries have become industrialized. In 1970 the US Congress passed the Clean Air Act and the Environmental Protection Agency was established.

D. The Promise and Perils of the Science-and-Technology Partnership: Genie Out of the Bottle

In the decades following the Second World War, public awareness and concern also became focused on the abysmal state of the rivers, lakes, and streams. At mid-century, human beings continued, as they had for thousands of years, to treat rivers as waste-disposal facilities. But by then populations were much larger and more urbanized. Industrialization, which involved an entire complex of new technologies, had resulted in increased quantities and types of wastes being dumped into rivers and other bodies of water. And the arrival of the powerful partnership, science-and-technology (discussed in Chapters I, II, and VII), had ushered in a whole new range of toxic waste products that were released into the environment. Most notable in this regard are the chemical industry and, to a lesser degree, the electrical and electronics industries, the development of which—unlike earlier technologies—was heavily dependent upon basic science.

The petroleum industry, one of the main suppliers of raw materials to the chemical industry, was itself launched on the basis of analyses of crude oil carried out by chemists. One of these was performed at the request of Samuel M. Kier (1813–74), an entrepreneur who was trying to find a use for the dark, odoriferous fluid that was a by-product of his salt-production operations. Wells drilled to obtain brine often yielded crude oil along with it. Kier had used it, instead of expensive whale oil, to illuminate his salt works at night, but burning it produced much foul-smelling, black smoke, rendering it unsuitable for most purposes. In the late 1840s, Kier even tried marketing it as a medicinal product. The chemist, James C. Booth (1810–88), advised Kier that petroleum had potential as an illuminate but that it would have to be distilled or refined. Beginning in 1850, Kier began experimenting with the distillation of crude oil. This resulted in a fuel oil that burned without objectionable smoke or stench, which he called "carbon oil." Right away, Kier entered into a partnership with John T. Kirkpatrick and they built a petroleum refinery in Pittsburgh, the first in the United States, but had to abandon the operation under pressure from fearful residents. In 1854, the Canadian doctor and geologist Abraham Gesner (1797–1864) was granted a patent for a distillation process that yielded kerosene (kerosine), a fuel oil suitable for lighting, from crude oil. He also found a way to obtain it from coal, hence the name "coal oil." In 1858, another Canadian, James Miller Williams (1818–90), sank an oil well and set up a refining operation near what is now Oil Springs, Ontario. However, his achievement was soon dwarfed by developments in the United States.

Once again, interest in petroleum was stimulated by an assessment by a chemist, Benjamin Silliman Jr. (1816–85) of Yale University, who carried out an analysis of "rock oil" (the literal meaning of *petr-oleum*) in the 1850s at the request of a group of entrepreneurs. His report affirmed emphatically that excellent illuminants and lubricants could be obtained from it. By the late 1850s, it was generally recognized that petroleum held great promise, but sources of it remained extremely limited—usually to seepage in certain locales. At the time, there was no way anyone could have known that there were pools of petroleum deep in the ground that could be tapped.

In 1858 James Townsend (1825–1901), a New Haven banker, gained control of the company which had sought Silliman's appraisal of petroleum and which had leased 100 acres of farm land and acquired the right to extract oil on an additional 12,000 acres near Titusville, Pennsylvania, where oil seepage was common. Townsend put Edwin L. Drake (1819–80) in charge of finding a way to obtain large quantities of oil from the properties and arranged financing for the undertaking. Although he had no geological expertise, Drake was convinced from the start that oil was trapped in rock cavities deep underground. He concluded that it should be possible to extract it by boring into the rock, just as was done to obtain brine for making salt. After various setbacks, Drake began drilling in the summer of 1859. At first he was able to achieve well-depths of only 16 feet or so. Mud and water rushed into the well shaft faster than it could be removed. In solving this problem, Drake may reasonably be said to have laid the foundation of the petroleum industry. Using steam power, he drove an iron pipe down the well and drilled down inside the pipe, while driving it deeper and deeper. (Later these pipes came to be referred to as "well casings.") Solid rock was encountered at a depth of 39 feet, oil at a little over 30 feet more. The oil rose up the well shaft to within a short distance of the surface. Drake fitted the well with a pump and began extracting oil from it at the rate of about 500 barrels a month. Remarkably, Drake did not seek a patent for his invention but actually showed others how to replicate his success. Right away, oil derricks began popping up like daffodils in the spring. As a result, the price of crude oil plummeted and the whale-oil industry collapsed, granting a (brief) reprieve to these immense marine mammals.

The petroleum industry developed during a time that has come to be recognized as a watershed in human economic affairs, a time when a new dimension was added to the picture: energy. As can be seen retrospectively, some of the elements of this development had been around for a long time. What was new, in addition to more of these elements, was the discovery and exploitation of relations between them. People have gathered fuel to feed fires to provide heat and light about as long as there have been people. But the invention of the steam engine made it possible to derive motive power from heat. Prior to that, the suggestion that such might be possible or that heat could be generated by motion would have lacked credibility. But by the middle of the nineteenth century, a transformation had taken place: railroad construction was booming, and (as discussed in Chapter VII) Rumford had demonstrated that motion could make water boil (1789). Then (in the 1840s) James P. Joule had nailed down (empirically) the equivalence relation between electrical energy (wattage) and heat (Joule's law) as well as that between mechanical energy and heat. At mid-century, scientists were in the process of elevating conservation of energy to its status as a basic law of nature. Rudolph Clausius (1822–88) is credited with being the first to formulate it in general terms (in the 1860s).

Of course, the existence of this theoretical basis did not translate immediately into practical applications or into changes in the way (most) people thought about matters to which it was relevant. Initially petroleum was sought not as an energy source, with all that implies, but simply as a fuel for a flame to provide illumination. That it might have other uses was certainly contemplated, but only in concrete terms at first. Its ability to supply motive power was realized only with the invention of the internal combustion engine and the automobile. The invention of the incandescent electric bulb (1879) and a stream of other electrical devices created a growing demand for electricity. In response, quite a variety of generating systems were developed. These utilized steam engines, water wheels, windmills, diesel engines, turbines of various sorts, and (eventually) nuclear reactors. The use of diesel engines for this purpose provided the petroleum industry with a new customer.

Electricity-generating facilities became progressively more sophisticated and, taken together, more reflective of the implications of the law of energy conservation—i.e., that there are numerous ways of producing electric power. Because of the desire for the benefits of electricity at the lowest possible cost under a variety of conditions, much attention was focused on how best to utilize the available options. For example, hydroelectric power was often used to meet peak demands because it was easier and more economical to increase or decrease the wattage thus produced than to change the output of a fossil fuel plant. Public awareness of such matters became even more heightened following the Second World War as the environmental costs of expanding and intensifying human technological activity pressed for attention and people sought "environmentally friendly" sources of energy. The involvement of the petroleum industry in the creation of environmental disasters is illustrated powerfully by what happened in Cleveland, Ohio and its environs.

Stoked by the growth of the chemical industry and the invention of the automobile, the petroleum industry soon accounted for a major fraction of economic activity in the United States. Initially, extracting, refining, and transporting petroleum and petroleum products were in the hands of a large number of small-scale operators, and conditions in the industry were volatile and chaotic. Convinced that consolidation would bring order and efficiency to the petroleum industry and (of course) profits to himself, John D. Rockefeller (1839–1937) set about acquiring one small petroleum business after another in the years following the Civil War. Because of its location and transportation advantages, he chose

Cleveland, Ohio to be the center of his oil-refining operations. To achieve his goal of monopolizing the petroleum industry, he formed an alliance with a few other businesspeople which was incorporated in 1870 under the name "Standard Oil." The corporation prospered handsomely during the ensuing decades as did the city of Cleveland, which continued to develop as a major center of the iron-and-steel and heavy-machinery industries. But this economic development was accompanied by environmental degradation that was, in the end, nothing less than spectacular.

Oil refineries, steel mills, factories, and warehouses lined the shores of the Cuyahoga River, which winds through Cleveland before emptying into Lake Erie. Industrial wastes dumped into the river and oil and its derivatives seeping into it from the refineries made it a poster child for polluted rivers. As incredible as it might seem, the Cuyahoga carried such a load of flammable fluids and other combustibles that it sometimes caught fire, putting bridges and ships in peril! The 1969 fire drew national attention—not because it was the worst ever (it wasn't), but almost certainly because of the growing public awareness of and concern with the deteriorating state of the environment. This movement gained additional impetus from the highly visible effects of pollution on Lake Erie, which included massive fish kills that left its shores littered with dead fish. Not only was Lake Erie inundated with industrial waste from cities along its shores and from rivers emptying into it, but it was also befouled by great quantities of raw sewage from the cities and towns and agricultural run-off laden with fertilizer and pesticides. The resulting high levels of plant nutrients in the lake (a process known as "eutrophication") fueled explosive growths of algae (known as an "algal bloom"). When algae die, they ordinarily sink to the bottom and rot, a process that removes dissolved oxygen from the water. If the concentration of dissolved oxygen in water falls below a certain level, fish suffocate. In the 1960s algal blooms in Lake Erie became so massive that wind-rows of rotting algae lined its shores, and in summer certain areas of it were carpeted with layers of algae up to two feet thick.

The "burning river" and the "dying lake" caught the media's attention and fired the public's determination that something be done. In short order, the US Congress passed the Clean Water Act (1972), which tightened restrictions on industrial waste disposal. In the same year Canada and the United States reached an accord, the Great Lakes Water Quality Agreement, which was focused primarily on reducing algal growth by controlling the amount of the plant nutrient phosphorus that was getting into the lakes. A revised version (1978) broadened the scope of the agreement to protecting the waters of the whole Great Lakes Basin Ecosystem. It emphasized preventing toxins from getting into the waters of the basin. The reductions, achieved in ensuing decades, in algal blooms and in the presence of toxins in the Great Lakes demonstrated that effective action could be taken to repair anthropogenic damage to the natural environment.

Another example of environmental destructiveness stemming from human inventiveness that attracted widespread attention in the 1960s is the havoc wreaked by the indiscriminate use of the insecticide **DDT** (dichloro-diphenyl-trichloro-ethane). Indeed, it would not be too far off the mark to say that *Silent Spring* (1962) by Rachael Carson (1907–64), which brought the ill-effects of DDT forcefully to the attention of the public, did more than anything else to launch environmentalism as a major cultural force. This episode also illustrates very effectively the promise and the perils of powerful new technologies made possible by the science-and-technology partnership.

Dichloro-diphenyl-trichloro-ethane was first synthesized by a German doctoral student in the late 1850s, but no notice was taken of it until 1935, when J.R. Geigy, a Swiss chemical dye company, set about finding a preparation that could be used to protect woolens from moths. It assigned this task to a team of chemists led by Paul Hermann Müller (1899–1965). After testing various chemicals on houseflies, they found that DDT was not only effective but also persistent (continued to work without reapplication for days and even weeks). Furthermore, the product did not irritate the skin of test subjects as did other preparations. The Geigy chemists then tested DDT on cockroaches, mosquitoes, and other pests with excellent results. In 1940, the chemical company began marketing the product as an insecticide, which it called "**Gesarol.**"

When the United States entered the Second World War in 1941, people in the Office of Scientific Research and Development were aware that disease, particularly those forms spread by insects such as lice and mosquitoes, was one of the most formidable challenges that troops in the field would face. Accordingly, a program to find ways to deal with disease-carrying insects was launched in 1942. When the First Marine Division had to be pulled from combat in the Pacific because over half of its men had fallen victim to malaria, this effort took on new urgency. Thousands of products were tested, but none proved satisfactory—until a sample of Gesarol arrived from Switzerland in the fall of 1942. In repeated tests under a variety of conditions, DDT demonstrated its suitability for the task. Subsequent field trials dramatically confirmed its capabilities, whether in powder form applied by hand-operated blowers or in kerosene solution sprayed from airplanes. Lessons thus learned made possible DDT's first major triumph: the suppression of an incipient typhus epidemic in Naples in the winter of 1943–44. When the city was liberated in October, people were living in deplorable conditions and as many as 90 percent of them were infested with body lice, the vehicle of the bacterium that causes typhus. So impressive was this result that the British and American armies made DDT a regular part of their arsenal. The insecticide also played a vital role when German concentration camps were liberated, as typhus was rampant in them. But as important as it was, the part played by DDT in meeting the typhus/lice challenge was not as significant as its contribution to the fight against mosquitoes and malaria. This becomes clear if one considers not only what happened to the First Marine Division and the role of DDT in the Pacific campaigns but also its contribution to the ongoing, worldwide fight against malaria. It has been estimated that by 1950, DDT had prevented five million deaths from that disease.

Prior to the summer of 1944, information about DDT was kept secret because of its role in the war effort. But as hostilities wound down the following summer, restrictions on civilian access to the chemical were lifted. Since it was known from preliminary testing that DDT was effective against a variety of insects, it was quickly utilized to solve many different kinds of domestic and agricultural as well as public-health related insect problems. Accordingly, DDT was sprayed wherever flies congregated or were not wanted: in homes, businesses and public buildings, and large, open spaces outdoors. Airplanes sprayed the insecticide on potato crops, soybeans, and—most of all—cotton crops, which had been devastated by the notorious boll weevil. Elm trees threatened by Dutch elm disease (caused by a fungus spread by bark beetles) were treated with DDT, as was a large forested area in New Brunswick infested with spruce budworm. DDT was also the centerpiece of a worldwide anti-malarial campaign. The entire island of Sardinia was subjected to four years of saturation spraying in a failed effort to eradicate mosquitoes.

In retrospect, this indiscriminate use of DDT may seem irresponsible. Even so, it exemplifies one of the basic ways that technological innovation takes place. As seen above, especially in Chapters V and VII, once a substance, mechanism, or procedure has been devised (for whatever purpose), someone can be counted on to try using it in some other way. Unfortunately, in the case of DDT, additional applications greatly magnified the harm it caused. Hence a second feature of technological innovation that is very much in evidence in the case of DDT is unanticipated side-effects. Since DDT is a broad-spectrum insecticide (it kills all kinds of insects), such unintended side effects are especially problematic. For example, declines—even decimations—of fish populations have resulted from aerial spraying because the insecticide wiped out the food supply of the fish, even when the insects the fish fed upon were not the target of the spraying! Another unintended side effect of the wholesale use of DDT is that it not only provides an opportunity for resistant strains of a pest to flourish but also facilitates their success by wiping out the organisms that prey upon the pest. This was forcefully demonstrated by the failed attempt to completely eradicate mosquitoes on Sardinia.

A third feature of technology exemplified by DDT is that it is often a double-edged sword: the very characteristics that make a technology so useful for one purpose are often the source of the problems it causes. The broad-spectrum nature of DDT makes it effective against fleas and mosquitoes as well as body lice (and thus an extremely valuable weapon against diseases spread by biting insects), but this very characteristic also makes it a menace to fish in affected bodies of water. The persistence of DDT is a great advantage in combating typhus. (It remains effective against lice for days and even weeks after application.) However, because it degrades slowly and is not soluble in water (and thus is not eliminated by the kidneys) but is soluble in fats and oils, DDT tends to accumulate in the bodies of animals that ingest it. This accumulation increases the higher up a creature is in the food chain. The result can be devastating for a species, depending upon the chemistry of its reproductive processes. Birds of prey seem to be especially vulnerable in this regard, as is evidenced by the steep decline in their populations in the years following the Second World War (when the use of DDT exploded) and the rebound in their numbers following the (1972) ban of its use.

One might think that once it becomes clear that the use of a substance is causing significant, perhaps irreversible harm to the environment, the appropriate action would be to prohibit it. Unfortunately, dealing with environmental problems is not always simply a matter of eliminating "harmful" substances. Certainly, the Lake Erie algal booms and fish kills were caused by phosphates that were getting into that body of water. But an influx of phosphates can be very beneficial. Phosphates are essential to the growth of algae, whether their presence is a problem or a key element of an ecosystem. Without the phosphates that are supplied continuously to the waters of the Georges and Grand Banks by upwelling, nutrient-rich ocean currents, the phytoplankton (tiny plant-like eukaryotic organisms that are the foundation of the food chain) would not flourish there and these massive shoals would never have been prime fishing sites. Phosphates are also needed by the plants that supply food for human beings and their animals.

The challenge at this point is to develop and implement ways of producing food that inflict less damage on the natural order than is being done currently. Even with the continued use of chemical fertilizers and pesticides, there are ways to reduce the percentage that gets into lakes, rivers, and streams. Some of these methods are simply good farming practices (contour plowing, strip-cropping, and crop rotation) that slow run-off, reduce erosion, and pay dividends to the farmer.

Others, such as leaving a buffer zone between cultivated land and rivers, lakes, and streams, have an economic cost. So does the abandonment of methods of applying insecticides, such as crop-spraying with airplanes, that deliver only a small fraction of the material to the intended target. Similarly, curtailing the use of broad-spectrum pesticides may require costly and time-consuming efforts to find alternatives with no guarantee of even partial success. The unavailability of acceptable replacements for environmentally-destructive products often triggers probing discussions and debates concerning the relative weights to be accorded the needs of human beings, on the one hand, and the integrity of the environment, on the other. One example is the controversy surrounding the flooding of the Hetch Hetchy Valley to create a water-supply for San Francisco, as discussed above.

Discussions and debates of this kind took place as the effort to ban DDT got underway. Although the insecticide was soon outlawed in industrialized countries where the threat posed by malaria was minimal, its value in the fight to suppress the disease where it was a major scourge weighed heavily against a total ban. After all, its use inside homes (for example, on mosquito netting) requires only a tiny fraction of the amount used in the applications that caused so much harm. The only problem with allowing such limited uses would seem to be ensuring that they remain thus limited, which is important, given the emergence of DDT-resistant strains of mosquitoes. For that reason, DDT could never be more than a small part of the answer to the malaria menace.

All of these cases and more contribute to the ongoing threat of anthropogenic (human-caused) climate change. The burning of fossil fuels continues to increase atmospheric concentration of carbon dioxide (CO_2), the primary cause of global warming, at an alarming rate. In 1940, approximately 5 billion metric tons of CO_2 were released into the atmosphere—by 2025, that figure had increased to approximately 38 billion metric tons. Clearing land for industrial development or agricultural use reduces trees that serve as CO_2 sinks—naturally reducing the amount of atmospheric CO_2. Though DDT is no longer used in the US, other pesticides continue to pollute the environment while paving the way for the increased grain production necessary for farming beef on an industrial scale, which contributes large quantities of the greenhouse gases methane and nitrous oxide to the atmosphere.

* * *

The snowballing effects of technology are not always positive. The exploitation of fossil fuels has increased humanity's ability to harness energy in ways undreamed of in earlier centuries—at the expense of the Earth's atmosphere.

Likewise, pesticides allow for controlling disease and enhancing agriculture, providing more food for Earth's growing population, but at the cost of polluting the environment and causing unintended harm to other ecosystems. Railroads across the Great Plains and refrigerated rail cars increased supplies of beef through much of the United States and lowered its price for consumers, but also increased pressure to exterminate buffalo that might block train routes. When contemplating the havoc wreaked, it is difficult to quarrel with the slogan, "We have met the enemy, and he is us," from the cartoon strip "Pogo" by Walt Kelly (1970), which appeared on a 1971 Earth Day poster. Clearly, this situation calls for human beings to stop and reflect upon the effects of their actions, to re-examine their priorities, and to re-think their place in the scheme of things. Issues of this kind, which involve ethics and values, that are raised by the impact of technology on the environment are addressed in the following chapter along with other normative questions raised by technological innovations.

Sources / Further Reading

Adler, Jonathan H. "The Fable of the Burning River, 45 years Later." *The WashingtonPost*, 22 June 2014. washingtonpost.com/news/volkokh-conspiracy/wp/2014/06/22/the-fable-of-the-burning-river-45-years-later/.

Bergman, Brian. "Buffalo Reborn." *Maclean's*, vol. 117, no.1, 16 February 2004, pp. 48–49.

Boorstin, Daniel. *The Discoverers: A History of Man's Search to Know His World and Himself*. Random House, 1963.

The Buffalo Hunters. Time-Life Books, 1993.

Burns, Bill. "Bison: Back from the Brink." *Beaver*, vol. 82, no. 5, October/November 2002, pp. 16–22.

Carson, Rachael. *Silent Spring: 40th Anniversary Edition*, with an essay by Edward O. Wilson. Houghton Mifflin, 2002.

Dav, Urja. "Edwin Drake and the Oil Well Drill Pipe." The Pennsylvania Center for the Book, Summer 2008. pabook.libraries.psu.edu/literary-cultural-heritage-map-pa/feature-articles/edwin-drake-and-oil-well-drill-pipe

Detweiler, Robert, et al., editors. *Environmental Decay in Its Historical Context*. Scott Foresman, 1973.

Diamond, Jared. *Collapse: How Societies Choose to Fail or Succeed*. Viking, 2005.

Dickey, Parke A. "The First Oil Well." *Oil Industry Centennial: Journal of Petroleum Technology*, vol. 59, no. 1, January 1959, pp. 14–25. https://www.onepetro.org/journal-paper/SPE-1195-G, doi:10.2118/1195-G

Evans. Harold. *They Made American: From the Steam Engine to the Search Engine: Two Centuries of Innovators*. Little, Brown and Company, 2004.

Fumento, Michael. *Science Under Siege: Balancing Technology and the Environment*. Morrow, 1993.

Geist, Valerius. *Buffalo Nation: History and Legend of the North American Bison*. Voyageur Press, 1996.

Gohau, Gabriel. *A History of Geology*. Revised and translated by Albert V. Carozzi and Marguerite Carozzi. Rutgers University Press, 1990.

Goudie, Andrew. *The Human Impact on the Natural Environment: Past, Present and Future*, 7th ed. Wiley-Blackwell, 2013.

Gugliotta, Guy. "The First Americans." *Smithsonian*, vol. 43, no. 10, February 2013, pp. 38–47.

Hamill, Sean D. "Unveiling a Museum, a Pennsylvania Town Remembers the Smog that Killed 20." *The New York Times*, 2 Nov. 2008, p. A22.

Hansen, Emma I., et al. *Memory and Vision: Arts, Cultures, and Lives of Plains Indian People*. University of Washington Press in articulation with Buffalo Bill Historical Center, 2007.

Hardin, Garrett. "The Tragedy of the Commons." *Science*, vol. 162, 13 December 1968, pp. 1243–48.

"History of the Great Lakes Water Quality Agreement." *Environment Canada*. International Joint Commission. ijc.org/en/what/glwqa-history

Hodges, Glenn. "First Americans." *National Geographic*, vol. 227, no. 1, January 2015, pp. 124–37.

"Human History of Extinction." *Science News*, vol. 186, no. 2, 26 July 2014, p. 5.

Jensen, Oliver. *The American Heritage History of Railroads in America*. American Heritage Publishing Co., Inc., 1975.

Kennedy, Robert C. "On This Day, June 6, 1874: The Last Buffalo." The New York Times Learning Network. *New York Times*, 6 June 2010. nytimes.com/learning/general/onthisday/harp/0606.html#explanation

Kirch, Patrick V. "Transported Landscapes." *Natural History*, vol. 91, no. 12, December 1980, pp. 32–35.

La Farge, Oliver. *A Pictorial History of the American Indian*. Crown Publishers, 1956.

Mann, Charles C. "Clovis Points." *Smithsonian*, vol. 44, no. 7, November 2013, pp. 58–59.

Martin, Albro. *Railroads Triumphant: The Growth, Rejection, and Rebirth of a Vital American Force*. Oxford University Press, 1992.

Martin, Paul S. "Pleistocene Overkill." *Natural History Magazine*, vol. 76, no. 10, December 1967, pp. 32–38.

McGinty, Jo Craven. "These Population Counts Aren't Always for the Birds." *The Wall Street Journal*, 13 September 2014, p. A2.

Mehringer, Peter J., Jr. "Weapons of Ancient Americans: Clovis Cache Found." *National Geographic*, vol. 174, no. 4, October 1988, pp. 500–03.

Nash, Roderick. *Wilderness and the American Mind.* Yale University Press, 1982.

Newton, Lisa H., et al. "Skunked: Crisis in the New England Fisheries." Chapter 4, *Watersheds 4: Ten Cases in Environmental Ethics*, 4th ed. Thomson Higher Education, 2006.

Parfit, Michael. "Diminishing Returns: Exploiting the Ocean's Bounty." *National Geographic Magazine*, vol. 188, no. 5, November 1995, pp. 2–37.

Pyne, Stephen J. "Indian Fires: The Fire Practices of North American Indians Transformed Large Areas from Forest to Grassland." *Natural History*, vol. 92, no. 2, February 1983, pp. 6–11.

Ray, Dixy Lee, with Liou Guzzo. *Trashing the Planet.* Regency Gateway, 1990.

"Refining Crude Oil: History, Process, and Products." Oil Region Alliance of Business, Industry & Tourism. www.oil150.com/files/refining-crude-oil-history,-process-and-products.pdf

Rotman, Michael. "Lake Erie." *Cleveland Historical.* Center for Public History + Digital Humanities. clevelandhistorical.org/items/show/58

Sandom, Christopher, et al. "Global Late Quaternary Mega-fauna Extinctions Linked to Humans, Not Climate Change." *Proceedings of the Royal Society B*, vol. 281, no. 1787. doi.org/10.1098/rspb.2013.3254

Schorger, A.W. *The Passenger Pigeon: Its History and Extinction.* Blackburn Press, 2004.

Slocum, Kenneth G. "The Dying Lake." *The Wall Street Journal*, 10 February 1969, p. 1.

The Smithsonian Book of Invention. Smithsonian Institution, 1978.

Stapleton, Darwin H. "The Short-Lived Miracle of DDT." *America Heritage of Invention & Technology*, vol. 15 no. 3, Winter 2000, pp. 34–41.

Taylor, Colin F. *Native American Life: The Family, the Hunt, Pastimes, and Ceremonies.* Smithmark Publishers, 1996.

—— and William C. Sturtevant, editors. *The Native Americans: The Indigenous People of North America.* Smithmark Publishers, 1991.

Utley, Robert Marshall, and Wilcomb E. Washburn. *The American Heritage History of the Indian Wars.* American Heritage Publishing Co., Inc., 1977.

Warner, William W. *Distant Waters: The Fate of the North Atlantic Fisherman.* Little, Brown and Company, 1977.

White, John H., Jr. *The American Railroad Freight Car: From the Wood-Car Era to the Coming of Steel.* Johns Hopkins University Press, 1993.

Wilson, Harold S. *Confederate Industry: Manufacturers and Quartermasters in the Civil War.* University Press of Mississippi, 2002.

Part Four: Transition to the Digital Age

Chapter X:
From Sheet Music to Streaming: A Perspective on the Emergence of the Electronic and Digital Age

As NOTED AT THE outset, most current discussions of technology are concerned with "high tech," computer and information systems along with associated technologies such as robotics. It has even been averred that a device or technique "is not technology at all" if it existed before the births of today's young adults. Given the rapidity and the accelerating rate with which technological innovation has been taking place, such an attitude is understandable. Indeed, a good case can be made that such an accelerating rate of innovation is unprecedented. But does it follow that "high tech" does not belong in the same class with older technologies, including those that provide its foundation? In a context in which such expressions as "cyberspace" and "the cloud" are ubiquitous, it is important to keep in mind that the content associated with these "spaces" resides in physical form in computers and servers. This means that deep down this content consists of patterns and processes of a physical nature. Only a little reflection makes it clear that computer and information technology satisfies the criteria set down in Chapter II, just as do the printing press and the (mechanical) cash register.

The dismissive attitude toward older technologies cited above also illustrates their tendency to become invisible as well the difficulty people have in appreciating the accomplishments they represent. If a device or technique was in general use when someone was very young and just learning "to talk," naturally he or she takes it for granted and gives little or no thought to it or to what life was like without it. If a person has no real awareness of the older technologies upon which an innovation is based, he or she cannot grasp the advance it makes into new territory nor accurately assess the contributions made by individual inventors and scientists. That this applies to "high tech" follows from the recognition that it can be characterized as a cluster of structures that have as their foundation electric and electronic systems established earlier. Although serious efforts were made as early as the seventeenth century to construct a mechanical calculator, the modern computer emerged only after it had been shown by George Boole (1815–64) that logical and mathematical operations could be carried out by manipulating binary digits, and (by John von Neumann and others) that those operations could be conducted by means of banks of many thousands of vacuum tubes. Computing, as now understood, got underway when a method was devised by Alan Turing (1912–54), among others, of specifying (in binary form) a sequence of operations that would carry out a given task. This activity came to be known as "**computer programming.**"

If one takes into account the necessity of delving into the nitty-gritty of practice when analyzing technology, it quickly becomes evident that an in-depth, systematic treatment of "high tech" would be an enormous undertaking. Adding to the challenge facing anyone contemplating such a project is the need, when gauging the significance of an invention, to take into consideration its effects on the way people live and on the way they think about themselves and the world. Furthermore, given that few areas of human concern or endeavor have escaped the influence of the new technology, just surveying the vast array of developments that it has spawned would require much more space than is available here—even without considering how they came about.

An insightful perspective on this complex of topics is afforded by focusing upon one strand of the human tapestry. By tracing the ways that technological innovations have shaped the way a major element of human culture has developed, it is possible to obtain a good sense of their importance. An excellent subject for a study of this kind is **music.** Electronic and digital technologies repeatedly have changed—indeed, transformed—the making and enjoyment of music. Of course this is by no means the first time that human beings' expanding technological repertoire has led to significant changes in these activities. As part of the starburst of aesthetic expression that occurred around 30,000 years ago, people began making whistles and flutes with finger holes from hollow bones using the specialized new stone tools that are characteristic of the period (as discussed in Chapter III). And doubtlessly there are hundreds, perhaps thousands, of other examples one could cite of the appearance of new musical instruments that embody the enhanced technological capabilities of their creators.

However, this chapter is not a historical survey of the invention of musical instruments together with the technologies that made them possible. Rather, the emphasis is upon those technological developments which have had a significant effect on a broad range of human activities and which underlie important changes in music-making and its enjoyment, such as those that provided the foundation for the electronic and digital age. As impressive as those innovations are, however, a good argument can be made that the most profound change in making and listening to music came about earlier when it first became possible to listen to music without performers being present, i.e., by means of **recordings.** This is now so taken for granted that it is all but impossible to imagine what a marvel it seemed at the time.

But sound recording was not the first new technology to bring about broadly inclusive changes in music. **Music notation** long preceded it. Following the usual pattern seen in the development of other complex technologies (such as agriculture, written language, counting systems, gunpowder, etc.), music notation emerged over a protracted period, beginning in the Middle Ages and lasting for centuries. This system of representing the sounds of a piece of music by written symbols provides a way for a musical work to be sung or played by someone who has never heard it. Paralleling the effects of writing on language and related matters (as seen in Chapter VI), musical notation fostered major changes in musical performance, the preservation and distribution of music, and its appreciation. It stores, standardizes, and makes more accessible the specification of the sounds that need to be made to produce a given piece of music. The invention of the **printing press** in the fifteenth century magnified the impact of musical notation. The mass production of **musical scores** by means of the printing press greatly increased access by the general population to musical works. In this context, the appetite that people have for music, including a growing desire for new compositions, together with copyright protections for such creations, have greatly increased the opportunities for sufficiently talented composers to devote themselves to work of this kind. But as significant as these developments are, they pale in comparison to those initiated by sound recording, itself a harbinger of more marvels to follow at a quickening pace.

A. Recording Sound

The first functional device for capturing and reproducing sound, invented by Thomas Alva Edison in 1877, is analogue in nature. His "**phonograph**" captures the acoustical properties of speech by linking them to the minute geometrical features of a line impressed by a stylus on the tinfoil surface of a small cylinder. This linkage is established by a diaphragm attached to the stylus which vibrates when struck by sound waves. Thus, when the cylinder is rotated by a crank, the line traced by the stylus on the tinfoil is given an extremely minute wavy character reflecting the characteristics of the sound. Screw threads on the shaft supporting the cylinder serve to shift its position with each rotation just enough that successive tracings do not overlap. To reproduce the recorded sound, the mouthpiece is replaced by a flared conical horn, the stylus returned to its original position, and the crank turned as before. But in this case, the stylus causes the diaphragm to vibrate and thus to produce sound waves. Edison attempted—unsuccessfully—to market the device as a replacement for written business and governmental communications. His phonograph did enjoy an initial, limited commercial success as a curiosity when coin-operated units with prerecorded content were placed where people congregate. Unfortunately, when the foil (and later the wax) cylinders were played repeatedly, they quickly wore out.

This problem was avoided in an alternative system, the "**gramophone**," which was patented by Emile Berliner (1851–1929) in 1887 (see Figure 10-1 below). Instead of cylinders, it employs flat discs made of shellac (a hard but brittle material), upon which spiral grooves are impressed by means of a master disc—later came polyvinyl chloride (PVC), giving rise to the term "vinyl" for a disc. Lacking any capacity to record sound, this system was designed for the mass production of discs with prerecorded content. Gramophones were offered to the public primarily as a means of listening to the music recorded on the discs. The sound "stored" on one of these "records" is reproduced in essentially the same way as with Edison's cylinders. A stylus ("needle") connected to a diaphragm causes it to vibrate as the stylus traces the spiral groove in the surface of the disc when it is rotated. (In time this came to be known as "playing a record" and all machines so used "phonographs.") In early models, discs were turned by means of a hand-wound spring mechanism like that used in clocks. Gradually, mechanisms were used to automate the process, allowing for different sizes of records that required playing at different rotations-per-minute (RPMs), including the LP ("long play") record, typically containing an entire album.

For the first time ever, people were able to listen to music at times and at places of their own choosing without performers being present. As noted above, this is

Figure 10-1. Gramophone. This illustration shows a gramophone with a turntable for a prerecorded disc and a hand-cranked rotational mechanism.

now so taken for granted that it is all but impossible to imagine what a marvel it seemed at the time. One result of the invention of sound recording, not to be discounted, was that a segment of the population that otherwise would never have heard such masterpieces as Sousa's marches and Rachmaninoff's piano concertos were able to do so.

The production, distribution, and enjoyment of musical recordings did not come into their own until the sound-capturing and re-creation processes were carried out electrically (or electronically). This capability, which brought about an enormous improvement in the sound quality of recordings, emerged over some fifty years. Achieving it depended upon several inventions being made and took place not in isolation, but in conjunction with innovations in a group of related technologies: the telegraph, the telephone, motion pictures, radio, PA systems, television, etc. Devices created to function in one of these areas were often put to use in the other areas as well. Furthermore, many of those who contributed to the creation or improvement of one of these technologies also had a hand in that of others. Perhaps most significantly, much of the work that produced these wondrous devices took place in that borderland between science and technology discussed in Chapter VII and was carried out by individuals who qualify both as scientists and as inventors.

An excellent example is the **microphone**, which plays a key role in the electrical recording of music. It was invented, as part of the **telephone**, to generate electrical signals that mirror the patterns of the sound waves that are produced by human speech. Although Alexander Graham Bell is generally credited with inventing the telephone, he was by no means alone in this pursuit. Furthermore, the type of instrument (containing liquid acid) used in his 1876 demonstrations suffered from such disadvantages that it never saw widespread use. A much superior device appeared only two years later. It was the initial version of the transmitter (or "microphone") that could be found behind the mouthpiece of telephones used during most of the twentieth century. By 1886 refinements made by Thomas Edison had brought to completion its basic design: a small cup of carbon granules covered by a diaphragm. Sound waves impinging upon the diaphragm put varying amounts of pressure on the carbon granules causing corresponding changes in electric currents flowing through them. Simpler, cheaper, more effective, and more durable than Bell's instrument, this microphone (transmitter) soon was found in millions of telephones. Further improvements to the microphone were prompted by the requirements of the nascent radio broadcasting industry.

By the time the race to devise a workable telephone was entering its final phases, methods of sending electrical signals over longer and longer distances were highly developed and increasingly sophisticated. Further improvements to the **telegraph** were already in the works as the great electrical/electronic revolution in communications began to pick up steam in the 1850s. Highly sought-after at the time was a way to send more than one message over a telegraph wire simultaneously. Among those attempting to solve this problem were Bell, Edison, and a number of others, including Elisha Gray (1835–1901), who would later become locked in a decade-long court battle with Bell over the patent for the telephone. The knowledge he gained in this endeavor, combined with his deep understanding of sound and speech (the focus of his professional career), put Bell in an excellent position to explore ways to carry out oral communication electrically.

Achieving this goal would not have been possible prior to the discovery that electricity and magnetism are closely related. In the 1820s, Hans Christian Oersted discovered that electric currents generate magnetic fields, and André-Marie Ampère (1775–1836) found that an electric current flowing in a loop of wire had the properties of a magnet. He further determined that the strength of these "**electromagnets**" was proportional to the intensity of the current flowing in the loops of wire. Thus, if there are fluctuations in the current flowing in such a coil, there will be analogous variations in the magnetic field generated. Consequently, an iron (or other ferrous metal) diaphragm or reed appropriately positioned with respect to a coil of wire can be made to vibrate in a way that mirrors the patterns of the sounds of human speech by sending an electric current that embodies those wave forms through the coil. This way of converting electrical signals into audible sounds is used in telephone "**receivers**" and in earphones.

B. Amplifying and Broadcasting Sound

During the early 1920s, electric currents that mirror sound patterns were put to work in many other ways, including the production of phonographic records. This application was made possible by the development of the **electronic amplifier**, by means of which electrical and electromagnetic signals once received can be sent on at greatly increased strengths. The perfection of the amplifier took years and involved the contributions of numerous very talented scientists and inventors. The breakthrough that led to the vast electronics industry, computers, the internet, etc. was the invention of the **vacuum tube**, specifically the **thermionic triode**, which was patented in 1907. Of course, more sophisticated, more powerful, and more specialized electronic devices were soon developed; but the invention of the vacuum tube marks the beginnings of the electronics industry and all that it gave rise to.

Basically, this device consists of an evacuated glass bulb with three metal elements projecting into it from the base, rather like the filaments of incandescent light bulbs. Indeed, one of the three, the "**heater**," is quite similar to such a filament, but its function is very different. When hot, the heater is surrounded by a cloud of free electrons. (Freeing electrons from the surfaces of heated metals is known as "thermionic emission.") If another of the metal elements, the "**plate**," is positively charged, these electrons will be attracted to it; and if the heater and the plate are connected through the base to a circuit that includes a power source, an electric current (stream of electrons) will flow from the heater to the plate, provided it is the positive terminal of the power source that is connected to the plate. No current will flow if the negative terminal is connected to the plate, as that would give it a negative charge, repelling the electrons from the heater. Vacuum tubes with only these first two elements (referred to as "thermionic diodes") are included in circuits when it is desirable to allow current to flow in only one direction. The third metal element of the thermionic triode, the "**grid**," is an open, grille-like structure located between the heater and the plate. Because of its open structure, the mere presence of the grid has only a negligible effect on heater-to-plate currents. But if it is negatively charged, electrons from the heater will be prevented (to various degrees) from reaching the plate. ("Like charges repel each other.") On the other hand, if the grid is positively charged, the flow of electrons from the heater to the plate will be enhanced. ("Unlike charges attract each other.") Thus, if the grid is included in a circuit similar to that involving the plate, the current in this grid circuit will affect the heater-to-plate current. The nature of that effect will depend upon the polarity of the grid current and the voltages involved. Most importantly, it was discovered that even if the grid current is very weak, small changes in it will produce large changes in the heater-to-plate current. This means that a weak signal arriving via the grid circuit will result in a strong signal being sent out on the plate circuit. That linkage makes it possible to use vacuum tubes to amplify electrical signals and electromagnetic ones as well. In the latter case, the current in the grid circuit is induced by the radio (electromagnetic) waves being picked up rather than being produced by a battery or generator, as would be the case if telephone messages were being amplified.

Right away, electronic amplification was put to work overcoming a long-standing limitation on voice transmission: distance. Thanks to this new technology, conversing by means of coast-to-coast telephone lines became possible in 1915. The first transatlantic radio-telephone voice message was received only a few months later. Electronic amplification not only makes it possible to receive radio signals from greater distances; in conjunction with the **loudspeaker**, it also enables people to listen to radio broadcasts without using earphones. As a result, listening to "**the radio**" ceased to be a solitary activity of a relatively small number of individuals and rapidly became much more: a common practice of individuals, families, various other groups, and entire populations. The new medium provided entertainment, information, and views on issues of all kinds. Commercial broadcasting began in 1920, and the number of radios in homes quickly grew from the thousands to the millions. By the 1930s, the number of homes equipped with radios was sufficiently large that Franklin Roosevelt (1882–1945) was able to make effective use of the medium in campaigning for office and subsequently in building support, in his "fireside chats," for his domestic programs and the war effort. In less than twenty years, radio became a core element in the culture of industrialized societies, so much so that it is now all but impossible to imagine life without it. A reminder of the importance that the new medium assumed was provided by the 2010 Academy-Award-winning movie, *The King's Speech*.

Along with several other types of programming, radio delivered musical performances to its listeners, greatly expanding (as had the phonograph) the segment of the population able to hear music as well as the range of works being heard and the amount of time that people were able to listen to music. During the same period in which the broadcasting industry was getting underway, the phonograph-record business was gearing up for its own golden age. By 1925, musical recordings were being made using electronically amplified signals from microphones to drive the electromagnets controlling the needles making the wavy, spiral grooves of the records. Further electrification in the 1930s led to the iconic "**record players**" of the mid-twentieth century. These machines feature electric motors for the turntables and magnets in the cartridges at the end of the tone arms which (as a result of the vibrations of the needles tracing the wavy grooves of the records) generate weak electric signals that are sufficiently strengthened by amplifiers to power loudspeakers.

Vacuum tubes can also be used as switches in electronic circuits. Indeed, they are, and are also called (electronic) "**valves**," notably in Great Britain. If the grid of a vacuum tube is positively charged, the switch is "closed." (Electrons are allowed to stream across the gap between the heater and the plate.) On the other hand, if the grid is negatively charged, the switch is "open." (Electrons are prevented from streaming across the gap between the heater and the plate.) The two states of an electronic switch, open or closed, correspond exactly to the two digits of a binary number system, 0 and 1. Thus the settings (open or closed) of sequences of these electronic switches can provide physical representations of binary numbers. This development, together with the work of Boole, Turing, and others, laid the foundation for the computer and information age.

While it might seem that the radio and record businesses were destined to be strictly competitive rivals, they actually developed a symbiotic relationship in which each benefited from the use that the other made of its products. Since playing records costs much less than broadcasting "live" performances, it is not surprising that programming of this kind became a major part of what radio stations had to offer. At the same time, record makers profited from the increased sales that resulted from their recordings being broadcast. Indeed, as things developed, all-out efforts to have records played "on the air" were not unusual. While the human appetite for new musical compositions had shown itself earlier (see above), the technological developments under discussion provided the basis for an unprecedentedly rapid and extensive introduction of new musical works into the public consciousness and public life. Soon rating systems were developed based on such factors as the number of times a given recording was aired by radio stations or sold by dealers during a given period. Rankings listing the "top ten," the "top twenty," or the "top forty" were announced on a weekly basis, as new releases competed for standing and established "hits" clung to another week in the top ranks whenever possible.

The availability of electrical recording devices, microphones, electronic amplifiers, and loudspeakers gave rise to an entire crop of new modes of aesthetic expression and appreciation. Among them were ways of singing that exploit qualities of the human voice that cannot be harnessed without the use of amplifying devices. A particularly notable example is "**crooning,**" a style of vocalizing that burst upon the scene in the late 1920s, just as radio broadcasting and electrical recording were picking up steam. By singing softly into a microphone, crooners can make themselves heard throughout large venues without having to "project." By doing so, they can vocalize in tones that, until the 1920s, were heard only in intimate settings. As a trailblazer for the new singing style, Rudy Vallée (1901–86) skyrocketed to prominence but encountered a storm of criticism from a variety of sources, notably from defenders of the established aesthetic standards for male vocalists. Since these norms were developed before amplification systems existed, they reflect qualities of vocalization associated with the ability to project, qualities exemplified in the powerful voices of such stars as Al Jolson (1886–1950) and Enrico Caruso (1873–1921). Despite opposition from this and other quarters, singers who adopted this new mode of expression quickly became immensely popular. These include Bing Crosby (1903–77), Frank Sinatra (1915–98), and Perry Como (1912–2001). The success of the 1986 musical production, *The Phantom of the Opera*, affords further evidence of the changes in aesthetic standards that were precipitated by electronics as well as the new opportunities for aesthetic expression provided by the technology. The eerily haunting number, "The Music of the Night," closes with the singer's voice fading away like a will-o'-the-wisp. An even more extreme example is found in **falsetto singing**, vocalizing in an artificially high register beyond the normal range of the performer. Since, with rare exceptions, such vocalization is low in volume, amplification is essential.

C. Case Study: The Electric Guitar and Rock Music

The new electrical and electronic technologies also made possible new ways of combining and modifying either actual sounds or oscillating electrical currents mirroring their vibratory patterns. An important example is the development of ways to incorporate the **guitar** into dance bands and other musical groups. The sticking point was that the traditional (acoustical) guitar was not loud enough to be heard when played in conjunction with brass instruments, saxophones, drums, pianos, etc. This limitation became a matter of concern during the "big band" era when it became desirable to include guitar-playing jazz musicians in the big bands. Unfortunately, the results of early attempts at amplification left much to be desired. Although guitars connected to amplifiers were commercially available in 1932, a truly satisfactory solution to the problems encountered along the way did not emerge until after the Second World War. Furthermore, success was achieved by making fundamental changes in the basic features of the guitar itself, alterations so radical that the "**electric guitar**" can reasonably be said to be a new musical instrument, distinct from the older version, which is now referred to as the "acoustical guitar."

The key invention that made the electric guitar possible is the electromagnetic "**pickup.**" This device was developed as a remedy for "**feedback**" problems encountered in efforts to amplify guitar sounds. Just as participants in radio call-in shows must sometimes be told to turn off their radios to prevent a disruptive feedback loop from being formed, so a way was needed to prevent amplified guitar music coming from loudspeakers from being picked up by the microphone or other input device used to capture the notes coming directly from the guitar. This problem became especially challenging at dance-hall volume.

Among those who played key roles in the development of the electric guitar is Les Paul (1915–2009, originally Lester William Polfuss), a consummate guitar player and innovator. Recognizing that the sound board and hollow body of the traditional guitar were unnecessary in an electrified version, he fashioned his own **solid-body** instrument from a piece of pine four-by-four to which he attached (on each side) a half of the body of an acoustical guitar (for appearances and ease of handling). Because of its construction, it came to be referred to as "the Log." During the 1930s and 1940s, Paul had a varied and successful career playing the guitar, often his own creation once it was completed. In 1945 he learned (by way of another guitarist) of a much-improved pickup devised by an electrical engineer who had been employed during the war making electronic equipment for the military. Paul then arranged for the engineer to make a pickup and amplifier for him.

The new electromagnetic pickups prevent feedback problems by sending electrical signals to the amplifier that mirror the vibratory patterns of the **guitar strings**—not those of sound waves produced by the guitar. This is accomplished by means of tiny magnets surrounded by coils of extremely fine wire. The pickups are mounted right under the steel and/or nickel guitar strings, so that the distance between them and the magnets in the pickups is very small. Since this distance is so short (and since magnetic attraction between two objects obeys an inverse square law with respect to the distance between them), small periodic variations in it result in small but significant cyclic changes in the attractive magnetic force acting between the magnets and the strings. The resulting vibratory motion of the magnets induces weak oscillating electric currents in the coils of wire surrounding them that mirror the vibratory motions of the guitar strings. (In accordance with Newton's Third Law of Motion, any force exerted upon the guitar strings by the magnets is accompanied by an equal force acting on them, but opposite in direction.)

When Les Paul first tried to interest the Gibson Guitar Corporation in manufacturing solid-body electric guitars, he got nowhere. But that changed abruptly at mid-century when another company began marketing just such a musical instrument. With input from Paul, Gibson lost no time in bringing out its own solid-body electric guitar, the highly regarded and much sought after Gibson Les Paul. The creator of the new guitar that prompted Gibson's course reversal was Clarence Leonidas ("Leo") Fender (1909–91), a self-taught electronics expert and business owner/operator who had never taken an electronics course or played a guitar. Yet he designed one of the most popular electric guitars ever made and founded the company that manufactured them. Fender's path to the Rock and Roll Hall of Fame began with a youthful fascination with electronics and a small radio-repair shop at his parents' California home. He continued to study electronics and to tinker with radios and electrical devices while majoring in accounting at Fullerton Junior College. Before landing a job as an accountant, he worked as a deliveryperson and bookkeeper. Even after securing a satisfactory position, he maintained an involvement with electronics, making amplifiers and public address systems for use in dancehalls and at sports events. During the mid-1930s, Fender, now married, held one accounting position after another; however, employment evaporated as the 1929 Depression affected the economy.

In what surely must have been an act of desperation, Fender opened a small retail and repair shop in Fullerton in the late 1930s. Called "Fender Radio Services," it offered a broad range of products and services focused on music and electrical and electronic devices. An important part of the business was the production, sale, rental, and repair of amplifiers and public address systems. This part of the enterprise brought Fender into contact with many musicians and band leaders, among them an inventor and musician named Clayton Orr ("Doc") Kauffman (1901–90). Kauffman not only played an electric Hawaiian (or lap-steel) guitar, but he had worked for the Rickenbacher (or Rickenbacker) company, which began manufacturing this type of musical instrument in 1931. (Another version was played in a horizontal position, supported by four legs. Known simply as a "steel guitar," it became popular in country and western bands.) Called the 'frying pan" because of its shape, this Richenbacher guitar has a small, round body and a long neck. Cast of aluminum in one solid piece, it has, as its most significant feature, an effective pickup. While at Richenbacher, Kaufman adapted his own guitars to amplification and experimented with ways to improve the design of musical instruments of this kind.

The confluence of the interests and backgrounds of Fender and Kauffman led to a collaborative effort to develop such a design and to the formation of a partnership, the K & F Manufacturing Corporation. In 1943, the two tinkerers constructed a crude solid-body guitar specifically to test different types of pickups, one of which was patented. The following year they applied for a patent for a lap-steel guitar featuring this type of pickup. They began production in November of 1945, marketing it in a kit that included an amplifier designed by Fender. He became convinced that manufacturing and marketing these instruments was more promising economically than his current business and was willing to put in the incredibly long hours of hard work this would entail. Kauffman, on the other hand, wanted time for family life and was more risk averse, so they parted ways in early 1946 but remained friends. Fender renamed the enterprise "Fender Manufacturing" and soon thereafter "Fender Electric Instrument Company."

Keeping the business afloat proved to be quite a challenge. Without Fender's focused dedication and dogged determination, it surely would have gone under. Other factors that contributed to its survival and eventual prosperity were the contributions of hard-working and talented people who joined the company as well as symbiotic business arrangements with other concerns. One of these was a tool-and-die company launched in Fullerton by two Second World War veterans who were looking for customers. Prior to their entering the picture, the production of guitars at Fender's shop was very much a handicraft operation, with each part being made by hand. But by this time, the company had reached the point that mass-production methods could be utilized profitably in the case of parts that were being turned out repeatedly. Furthermore, its future as an economically successful enterprise depended upon doing so. That is where the veterans' company stepped in, making specialized tools and dies whereby metal components for the guitars could be stamped out on punch presses. Thus began a mutually profitable relationship that was to last for decades. As the demand for Fender's guitars increased, the tool-and-die company developed more complex and sophisticated high-production techniques to manufacture them.

Another important contributor to the survival and success of the Fender Company was Don Randall (1917–2008), general manager of Radio & Television Equipment Company, of which Fender was both a customer and a supplier. As the exclusive distributor of K & F products (from early 1946),

Radio-Tel (as it was called) mounted a concerted effort to persuade dealers to stock them. Later on, Randall performed an even more vital function. By attending meetings of the relevant professional associations and talking with others in the trade, he was able to keep Fender informed regarding other manufacturers' latest offerings and the features that seemed to be attracting buyers. Randall even took it upon himself to convince a reluctant Fender to include some of these in the instruments he made. Such actions on Randall's part were especially important in the late 1940s when Fender turned his attention to developing an electric version of the Spanish or "standard" guitar, as it was known at the time.

Although it had been less popular than the Hawaiian version earlier in the century, the Spanish guitar had moved ahead rapidly. (By the 1970s it had so completely eclipsed the Hawaiian instrument that anyone speaking of a "guitar" could be assumed to have the Spanish kind in mind.) Because of high demand, Spanish (acoustical) guitars equipped with pickups and amplifiers had been available since the 1930s; however, as noted above, they had significant limitations. During the 1940s, a number of individuals and companies experimented with solid-body electric (Spanish) guitars in a variety of forms. At some point, Fender decided to develop and produce an instrument of this kind. Although it is not known just when or under what circumstances he came to this decision (he wasn't one to share his thoughts on such matters), it is no surprise that he took this route. He was an entrepreneur able to spot opportunities and willing to take risks as well as work incredibly hard in order to succeed. Furthermore, he was aware of the problems plaguing electrified acoustical guitars and of the need to solve them. And he had both experience and production capabilities for making solid-body guitars, which embody a solution to the feedback problem.

Although not a musician, Fender had a very definite idea of the sound quality he wished to achieve: the "electric sound," the pure sound of guitar strings being plucked. His goal was to reproduce it, uncontaminated by overtones and resonances deriving from the hollow body of the acoustical guitar and free of distortions introduced by pickups and amplifiers. He was also eager to produce an instrument that musicians would like and would want to possess. Accordingly, he canvassed local guitar players about the design of the prototypes. He placed the instruments in their hands for use in performances and experimented with the configuration of the pickups and amplifiers, even interrupting shows to make adjustments and to suggest changes in the settings of the controls on the guitars. The first Fender solid-body electric (standard) guitar, the "Esquire," was offered for sale in the spring of 1950, but initial models were plagued by problems, and sales were few. While these problems were being addressed, Fender set to work on a new design that included—at Randall's insistence—two pickups (instead of

Figure 10-2. The Broadcaster and Telecaster. This photograph shows the Fender Broadcaster and a version of the Telecaster available in 1953. Image credit: John Seb Barber. CC BY 2.0

one) and a steel stiffening rod in the neck. Initially called the "Broadcaster," the name of this electric guitar was changed to "Telecaster" to avoid a copyright conflict (see Figure 10-2). It and the modified Esquire were offered for sale in the spring of 1951.

As experienced businessmen, Fender and Randall knew that, contrary to the popular saying about a better mousetrap, a new product has to be brought to the attention of potential customers and they must be convinced of its value in order for it is to succeed. Here again, Randall played the vital role of promoting the new instrument. He and Fender took one of the early Telecasters to a nightclub where a prominent guitarist, Jimmy Bryant (1925–80), known for his extreme playing style, was performing. His experiments with the new Fender electric guitar quickly captured the attention of everyone present. Dancers and the other band members gathered around to listen. This example illustrates a central element of Randall's exceedingly successful promotional campaign strategy: place the instrument in the hands of highly visible guitar players so that it can be seen and heard as well as receive an implied—if not explicit—endorsement. Photographs of musicians holding Telecasters were a staple of Fender advertisements. Not only was the Fender solid-body electric guitar the first such instrument to be put on the market, it also achieved commercial success, becoming one of the most popular instruments ever.

The solid-body electric guitar was a success not only because it avoids problems encountered with electrified acoustical guitars but also because it enables performers to make novel uses of sound qualities and patterns that resonated with audiences. Arguably, the most profound change ushered in by the instrument is the acceptance of **loudness** as a desirable quality, an aesthetic norm. Of course, volume has long been an important dimension of musical expression, from the fortissimo of rumbling drums, clashing cymbals, blaring trumpets, and thundering pipe-organs to the pianissimo of strings and woodwinds. In contrast, a characteristic feature of **rock music**, for which the basic instrument is the electric guitar, is sustained, extreme loudness such as that produced in live concerts by powerful amplifiers and numerous loudspeakers. This open recognition of the essential role of electronics in the production of rock music is the very antithesis of the earlier practice of concealing loudspeakers in an effort to prevent listeners from realizing that what they were hearing was a reproduction of a human voice rather than an actual one. Thus, not only did the electric guitar give rise to a new—and very popular—musical genre, it also paved the way for a shift in aesthetic standards.

D. Magnetic Tape and Sound Engineers

The invention of the electric guitar was but the first step in the development of a cluster of devices and techniques used in a new approach to making music and in delivering it to listeners. What listeners hear in this context is the end product of an entire array of complex technological processes, sounds that can be heard only from loudspeakers or earphones. One of the methods thus employed is "**over-dubbing**" or "sound-on-sound," in which either a live performance is combined with a prerecorded one or two recordings (made separately) are combined. For example, one might make a recording of a musician performing while a record is being played (as in someone's karaoke rendition). Although some experimental work was done as early as 1930 using disc recording technology, it was only after the invention of recording on **magnetic tape** that over-dubbing came into its own. Not only was Les Paul among those who engaged in these early efforts, he was also a trailblazer in exploiting the opportunities provided by the **tape recorder**, which became available in North America in the late 1940s.

The magnetic tape recorder was invented by German engineers during the Second World War. By 1947, a small American company (Ampex) had developed an improved version and was starting to give demonstrations to professional audio groups. The close resemblance of these tape recordings to live performances made a powerful impression on audiences. Word of the device reached Bing Crosby, who had wanted for some time to pre-record his syndicated radio show but was prevented from doing so by the relatively poor quality of current recordings. Following a private exhibition, Crosby arranged for a test recording to be made of one of his programs. This trial was so successful that he immediately contracted to have the rest of the series recorded. Recognizing the potential of the new machine and eager to exploit it, Crosby threw his support behind the small company, investing his own money in it. Not only was Crosby the first American musical star to tape-record radio broadcasts, he led the way in taking advantage of the opportunities afforded by the new technology for improving radio programming. His tape-recorded broadcasts had an unprecedented smoothness and consistency. This result was achieved by assiduous **editing** thanks to the tape recorder.

In 1948 Crosby arranged for Les Paul to receive one of the first Ampex Model 200A tape recorders, which greatly facilitates over-dubbing. Crosby employed this technique in recording "How High the Moon," widely regarded as a watershed in popular music. It occupied the top spot on the *Billboard* singles chart for nine weeks in 1951. To make this recording, Les Paul repeatedly over-dubbed different instrumental and vocal parts, all of them consisting of his guitar playing and Mary Ford's singing. A more dramatic example is a 1991 Grammy-award-winning virtual duet, "Unforgettable," featuring Natalie Cole (1950–2015) and her father, Nat "King" Cole (1919–65).

The tape recorder not only facilitates over-dubbing; it gave rise to an entire crop of ways to manipulate sound recordings. Among them is **splicing**, joining together segments of different recorded performances. For example, by combining the best parts of several takes of the same musicians performing a work, it is possible to produce a recording superior to any that can be obtained from a single one of their live performances. Of course, since its beginnings the motion picture industry has been stringing together the different "takes" that make up a movie. In that case, however, the splicing was a physical process, piecing together actual lengths of film. In contrast, once tape recorders were available, parts of performances recorded on magnetic tape could be spliced into others electronically. **Speeding up or slowing down** the tape drive can be used to effect changes in tone quality, pitch, and rhythm (as was done to produce the chipmunk voices of Disney's Chip 'n' Dale). Recordings thus created can be combined with others to achieve novel effects. Still another technique is "**slapback**," which consists of combining two essentially identical copies of the same recording—but with one of them slightly delayed (typically by no more than a fourth of a second). The result is a recording that mimics the powerful, reverberant sounds heard in opera houses and similar venues.

The introduction of **multi-track tape recording** made even more options available. A multi-track tape has two or more strips ("tracks") of easily magnetized material along it. Thus, the sound captured by two or more microphones can be recorded simultaneously. And since they are recorded on the same tape, they are perfectly synchronized. Multi-track recording first became important in making stereophonic ("stereo") recordings. By using two microphones placed at different locations, recording (on separate tracks) what these microphones pick up, and sending electronic signals from each of these to a separate speaker, it is possible to produce recorded music with much more of the fullness and depth of a live concert. The two-track tape recorder also greatly facilitates experimentation with different sound combinations. It made it much easier for Les Paul, who was at the forefront of those thus engaged, to try different combinations of sounds in his quests for something distinctive and to form and compare different pairings of his own renditions. With a minimum of effort, he was able to listen to a pre-recorded part (performed by him) while recording another part (on the second track), just as each vocalist performing a duet listens to the other as they sing. This process can be repeated as many times as are desired.

Yet more ways to manipulate sound became available when the three-track tape recorder was introduced. Among many other possible arrangements, one track can be reserved for a lead singer or instrumentalist while the other two are devoted to backing music, allowing for it to be recorded stereophonically or to be produced by over-dubbing separate parts. These devices were widely used to make popular-music recordings well into the 1960s. By the end of that decade, four-track tape recorders, which provided much greater flexibility in recording and over-dubbing, had become the norm. Ever in the lead in the pursuit of new, complex sounds, Les Paul bought the first Ampex eight-track tape recorder.

In simplest terms, music traditionally has been understood to be provided to listeners by composers and performers. But as seen above, for some time now the contents of musical recordings have been pulled together and shaped by sound engineers and the producers of audio wares. While members of this new group do bring technical and economic perspectives to the table, they also participate in the shaping of aesthetic expression. This technological and aesthetic melding is epitomized by the career of Les Paul, who was a guitarist, composer, inventor, and his own **sound engineer**. While he was able to "do it all," bringing a sound engineer on board soon became a necessity for nearly anyone wishing to avail themselves of the new sound technology. The central role into which producers stepped is exemplified by the part played by George Martin (1926–2016), the "fifth Beatle," in creating the tracks featured on the album *Sgt. Pepper's Lonely Hearts Club Band*. Using techniques such as those discussed above, Martin and the Beatles worked together in a recording studio to make "Strawberry Fields Forever" and "Penny Lane," released in 1967. Numbers such as these cannot, of course, be replicated live onstage.

The addition of a new member to the traditional music triad (composer, performer, and listener) has not been confined to the recording studio. Musical groups using microphones and pickups when performing live must have someone to operate the **mixing board** or **console**, a piece of equipment into which electronic signals from these input devices are fed and from which signals go out to the loudspeakers, to the headset of the operator, and to the monitor speakers for the performers. Because of the earphones and the controls at his or her disposal, the sound engineer can play a major role in shaping the sounds that reach the ears of the audience. He or she is able to control, among other things, how much the incoming signals are amplified so that the desired balance and dynamics among the instrumental and vocal parts can be achieved. The operator can also decide which loudspeaker(s) will receive the signals from a given source at a given time as well as exert some influence over tone quality.

In comparison to the importance of the innovations in the production of music made possible by the tape recorder, the impact of the invention on music's accessibility to listeners in the 1950s was fairly modest. This changed dramatically with the development of miniaturized electronic circuitry, as discussed in the next section.

E. From Transistors and Integrated Circuits to Microcomputers

The mid-twentieth century witnessed the invention of a series of devices that greatly expanded both the range of musical performances that a typical person is able to enjoy and the range of circumstances under which he or she is able to do so. The key to these achievements was the **miniaturization** of electronic circuitry, made possible by the invention of the transistor and the integrated circuit. In an effort to find a less bulky, more reliable, and more energy-efficient alternative to the vacuum tube, a research team consisting of three physicists was formed at Bell Telephone Laboratories in the 1930s. Their investigations revealed that the conductive properties of **semiconductors** (substances such as silicon or germanium which are neither good conductors nor good insulators) can be altered dramatically by the presence of certain impurities. The Bell Labs physicists also discovered that these materials conduct electricity in two fundamentally different ways, depending upon which of two types of impurities are present. Moreover, they learned to introduce impurities into specific tiny regions (or layers) of slivers of semiconductor crystals. (This procedure is called "**doping**.") They even learned to introduce different types of impurities into tiny regions adjacent to each other. Most importantly, they were able to explore the properties of pairs of adjacent, but differently-doped tiny regions (or layers) by including them in electronic circuits. This led to the discovery that an electric current will flow in only one direction across the boundary between two such differently-doped tiny regions (or layers). As a consequence, these tiny systems can be used as "**rectifiers**" (devices to change alternating current into direct current) or as valves in electronic circuits, both very important elements of such systems. Previously these functions had been accomplished by means of thermionic diodes (described above).

Investigations of the properties of electronic systems that involve tiny semiconductor crystals with *three* doped regions (or layers), one of them doped differently (in the above sense) from the other two and located between them, led to another important discovery: in a certain configuration of two circuits, each linking two of the three regions (or layers) of such a crystal, a weak signal in one of the circuits results in a much stronger signal in the other. The upshot is that such contrivances can be used as amplifiers. By December 1947, the Bell Labs team was able to stage a demonstration of the amplification capabilities of this new device.

Christened the "**transistor**," it was first put to practical use in hearing aids, in which compactness is key. These were soon followed by the **pocket-size transistor radio**. First offered for sale in 1954, this invention exploits the reduced requirements of transistors for both space and energy (see Figure 10-3 below). The pocket transistor gave the individual greater control over what he or she listened to and increased the range of circumstances under which he or she could enjoy doing so. Individuals gained even further control over both the content and the circumstances of their listening when the compact cassette player became available in 1962. Unlike the phonograph which cannot be moved about while playing without interfering with its operation and risking damage to the record, the cassette player makes it possible for people to enjoy music of their own choosing almost anywhere. (The supply reel and the take-up reel holding the magnetic tape are enclosed in a protective plastic case that keeps the tape aligned with the pickup.) Although television is generally credited with bringing the golden age of radio to an end, the arrival of personal music devices may be regarded as relegating to history books the image of families or other groups gathering around the radio to listen to programs.

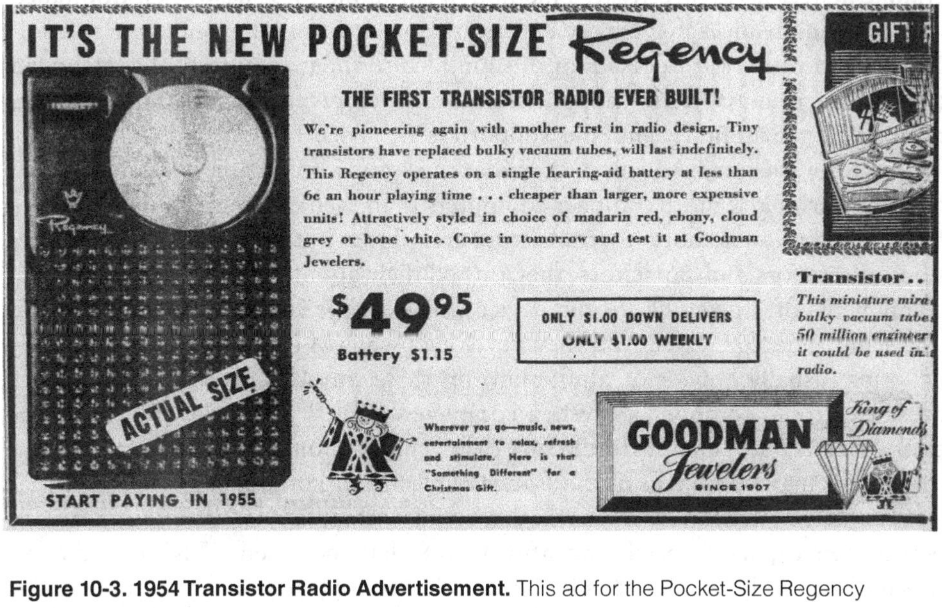

Figure 10-3. 1954 Transistor Radio Advertisement. This ad for the Pocket-Size Regency Transistor Radio showcases the remarkably small size at which a radio could be produced using transistor technology. Image credit: Joe Haupt. CC BY SA 2.0.

Another step forward in the miniaturization of electronic devices took place during this same period with the introduction of "**printed circuits.**" As circuits became more complex, so did the tangle of insulated wires linking their components and the more time it took to solder the connections. The solution to this problem involves dispensing with essentially all insulated wires and replacing them with a geometrical network of narrow strips of conducting material (usually copper) lying on the surface of an insulating board. A network of this kind is produced from a film of copper (or other conducting substance) covering the surface of an insulating board. Etching is used to remove all of the copper film *except* the narrow strips making up the desired network. This result is achieved by "printing" a photographic image of the planned circuitry onto the metal film—not with ink but with an acid-resistant compound which prevents the narrow strips from being removed chemically when the circuit board is subjected to an acid bath. Since electronic devices thus produced cost less and require even less space than earlier models, music became more accessible to yet more individuals under an even greater variety of circumstances.

The photographic method used to produce printed circuits provides an opportunity for an even more dramatic reduction in the size of electronic circuitry. A simple application of the principles involved in enlarging or shrinking photographic images leads directly to a method of shrinking these networks. Combined with further advances made by scientists and inventors in semiconductor technology, the utilization of processes of this kind led to the invention of the **integrated circuit**. The primary initial impetus for the development of integrated circuits came from the military and the supporters of space exploration, both vitally interested in rocketry. Advances in guidance systems, for example, require much more powerful and more compact computers than then existed. Once researchers had learned to make not only transistors and diodes but also semiconductor versions of resistors and capacitors, they turned their attention to fabricating all the components of a given electronic device (for example, an amplifier) on a single tiny semiconductor wafer. And they developed ways of locating miniscule metallic strips (usually copper or aluminum) on these tiny bits of crystal to connect those electronic components. When completed, such a system of electronic circuitry is called an "integrated circuit" or a "**chip.**" About the size of a small pea, these chips may house tens of thousands of electronic components.

Even more remarkable than the size of microchips is the accelerating rate at which their capacity, speed, and affordability has increased. This phenomenon was noted as early as the 1960s by Gordon E. Moore (1929–2023), who later co-founded the semiconductor company 'Intel' that eventually would dominate the microprocessor market, and others. In 1965 he published a graph in the magazine

Electronics depicting the growth (over the preceding 15 years) in the number of components (e.g., transistors) that an integrated circuit could be made to encompass. This graph includes an extrapolation for the next ten years. When it became evident (in 1975) that this exponential pattern of growth was continuing, the practice of referring to it as "Moore's Law" became established. Amazingly, a doubling time of about two years was maintained in ensuing decades, no doubt in part because it was self-filling. Very much in evidence here is the snowballing effect of technology supercharged with an awareness that innovation is possible and the expectation that it will happen.

Following these initial successes, **microelectronics** developed rapidly. Computers built around semiconductor chips are much reduced in size, faster, more reliable, more energy efficient, and cheaper to make than those available previously. They were first put to work in guided missiles and satellites, where weight and space are at a premium. They assumed an important role in the step-by-step human venture beyond Earth's atmosphere in the 1960s that culminated in a landing on the surface of the Moon. In the following decade, the transformative power of microelectronics began to make itself felt in the consumer-products area, beginning with the hand-held calculator. Of much more significance is the appearance of the **microcomputer**—not so much for what each individual unit was capable of doing but because of the possibilities that arise when large numbers of people have access to them. These "third generation" machines are much smaller and much more inexpensive than those of the previous generation which (by exploiting the advantages of transistors) had brought computer technology out of the research lab into the broader economy. But because of their price-tag, computers remained out of the reach of most people—until 1977, when Steve Jobs (1955–2011) and Steve Wozniac (b. 1950) brought out the Apple II.

In retrospect, the impact of the availability of these relatively inexpensive machines was clearly minimal until sophisticated programs were also readily available. These were not long in coming. In 1979 a spread-sheet program for the Apple II (dubbed "VisiCalc") was offered for sale at $250 a copy. It was created by a graduate student at the Harvard Business School in partnership with an MIT graduate. Within four years, nearly three quarters of a million copies were sold. Thus was born another addition to the economies of industrialized countries, the production of "**software**" (programs to be run on computers). Prior to the invention of the microcomputer, computing at businesses and other organizations was carried out on centralized ("main frame") systems, large numbers of which were manufactured by International Business Machines Corporation (IBM). Sensing both an opportunity and a threat, IBM quickly brought out its own small, personal computer (the "PC") in 1981. It was built around an Intel microprocessor and provided with an operating system from a small company founded by Bill Gates (b. 1955) and Paul Allen (1953–2018). Because it produced software for microcomputers, they called the company "Microsoft."

F. The Digitization of Sound

It would be hard to overstate the significance of the creation of the computer-and-information system that now spans the globe. Few, if any, areas of human endeavor have escaped its influence. Once again, both the way music is given the form that it takes and the way(s) it is delivered to the listener have changed radically. Furthermore, these continue to morph right before one's eyes. The key element in this development is the **digitization of sound**, i.e., its capture, storage, and reproduction by means of numbers rather than by means of analogous physical processes, such as wavy grooves in the surface of a rotating disc or variable patterns of magnetization in a coating on a plastic tape. The first successful system for recording music in digital form was created in 1972, when computing was still a main-frame operation. It was developed, beginning around 1965, in an effort to avoid the noise and distortion that accumulate with each step in analogue processes. Creating a digital record of an electronic signal coming from a microphone or pickup involves measuring its amplitude thousands of times per second and recording the results in numeric form. The frequency with which this sampling is carried out depends upon the quality that is desired in the ultimate product. The great advantage of a digital recording is that it can be accessed to feed an amplifier/loudspeaker or to make copies without distortion or adulteration as long as the numeric values can be read. Accordingly, as one might expect, the method was first put to use in making master recordings on magnetic tape. But copies for consumers continued to be offered in the vinyl-disc (LP) format. This would not change until 1982.

Meanwhile a radically different way not only of recording sound but also of storing information of all sorts was being developed. As is the case with many inventions, this one incorporates another new device, the **laser**, from the acronym for "light amplification by stimulated emission of radiation." Lasers operate by putting a population of atoms into an excited state and inducing them to radiate that energy in the form of a sharply focused beam of coherent light (i.e., light of a single frequency or color and directed very precisely in a single direction). Of special importance to the new method of storing and retrieving information is the invention in 1962 of the **semiconductor laser**. Since, like other semiconductor devices, these lasers are so much more compact, cheaper to make, and energy efficient than other types, they are especially suited to the proposal put forward by James T. Russell (b. 1931) in 1965: that one could use a laser beam to "read" a certain kind of physical record without altering it in any way.

Undeterred by skeptics, Russell continued to work on perfecting the invention during the 1970s. Others made contributions as well. The type of physical records that eventually emerged consists of a spiral track on a flat disc similar to that of a phonograph record but differing in two especially important respects. These tracks are incredibly narrow (only some 500 nm wide) and close together (separated by about 1.5 μm). Secondly, the information stored on them is binary in form. It is embodied in the presence or absence (at a given point along the track) of a tiny trench (or "pit," as they are called). But whether or not there is a pit at a given location is *not* what the reflected laser light reveals as the disc is rotated. Rather, it is whether or not one of these tiny trenches begins or ends there. If one does, the result is a change in the reflected laser light which can be detected by a photoelectric cell. The presence of such a boundary is registered as a "one"; the absence thereof is registered as a "zero." The pattern in which these "1s" and "0s" occur can be translated by a computer into an electronic signal that, when sent to an amplifier/loudspeaker, produces a reproduction of the sounds recorded on the disc—reversing the process by which the recording was made.

Pre-recorded **compact audio discs** of this kind (later referred to as "CDs") along with CD players first went on sale in the fall of 1982. Although the CD players were very expensive initially, the new system quickly gained popularity because it offered superior sound quality, durability, and convenience. Consequently, older music-reproduction methods were soon of little more than historical interest. As is often the case with new devices, the price of CD players soon began to decline, making it possible for more and more people to afford them. Among the factors that contribute to such outcomes are the amortization of start-up costs and the development of more efficient production techniques. (Of course, whether and to what extent such price reductions take place depends on a number of conditions.) Within a few years, CD players were selling at a fraction of the original cost and CDs were dominating the market. Furthermore, continued research found ways to cram more and more information onto these discs, so much so that it became practical to put the entire content of a set of encyclopedias on a single CD-ROM ("read only memory" disc). At one point, compact discs were available that could hold almost 50 times as much information as the hard drive of a typical computer. These were used extensively when space was needed for large amounts of digital data or for very large programs. But accelerating increases in the capacity, speed, and affordability of micro-chips continued unabated right into the first decade of the twenty-first century. Thus, despite advances made in the data-storage capabilities and versatility of CDs, their use as repositories for numerical data ceased to be significant as the new century got underway.

Once sound is digitized, it can be stored or manipulated as well as conveyed to or retrieved from another site in the same manner as other kinds of numerical data. Seemingly limitless is the range of possibilities thus introduced. Not only can the pitches and rhythms of recorded performances be altered, but short segments of one recording can be intermingled with segments of another, vocalizations from one recording can be grafted onto the instrumentation of another, etc. However, severe limitations on the electronic transmission of digitized music remained until the early 1990s because of the enormous amount of data involved. Fortunately, the quantity of numerical data required for satisfactory sound reproduction can be reduced by taking advantage of the limitations of human hearing. In order to make a recording that most people cannot distinguish from the original, it is not necessary to duplicate every acoustical aspect of it. Audio data files can be scaled down by eliminating redundancy and inessential information. (In particular, soft sounds that are "masked" by very loud ones need not be recorded with the same fidelity as others.)

The impetus to produce a computer program to accomplish this objective did not come from the music business, however. It came from the motion picture industry. By the late 1980s the desirability of digitizing films (including their soundtracks) was widely recognized, and efforts to realize this goal were well underway. In 1988 a meeting of engineers and executives (who came to be referred to as the "Moving Picture Experts Group") was convened for the purpose of finding a way to compress the immense video and audio files of a movie sufficiently to permit them to be stored on a computer and sent over the internet. An important part of the task was to establish standards for the compression of audio-visual data files. The project, in which a number of organizations cooperated, came to fruition in the mid-1990s in the form of MPEG-2 Audio Layer 3. This software, "**MP3**" as it is commonly called, reduces the size of CD-quality audio files by a factor of 10 or more. Toward the end of the decade, the software had become available on the web and music aficionados began converting CD audio files into MP3 versions and posting them on the web where they could be freely copied by others.

G. Satellites, Optical-Fiber Cables, and Cellular Networks

A prominent feature of the period extending from the present back to the end of the Second World War is the rapid rate of innovation in communications technology. Arguably, satellites "parked" in **geosynchronous orbits** constitute the most dramatic departure from earlier means of getting messages from one place to another. These circle the Earth at a distance such that the speed necessary to keep them in orbit is the same as that which keeps them over the same spot on the Earth's surface. Three of them strategically placed make it possible for electromagnetic transmissions to reach around the globe. Communications satellites were first launched into space in the 1960s as the space age got underway. Since then the number, types, and uses of these devices have mushroomed. When teamed up with cable television systems, they brought about fundamental changes in TV programming.

First put into operation around 1950 simply as group or community antenna services, **cable TV** systems were restricted (by a variety of factors) for almost a quarter of a century to relaying signals received "over-the air" from broadcasting stations. But by the mid-1970s these barriers were gone. Since cable systems bypass the interference problems faced by broadcasting stations, their subscribers are able to receive the offerings of an almost unlimited number of TV stations located anywhere on the globe (thanks to communication satellites). They are also able to enjoy programming available only on cable including movies, specialized channels, and short features. By providing a theater for short films, cable TV created yet another opportunity for that ancient partnership, music and spectacle, to work its magic. The power of this partnership was recognized by Aristotle and is perhaps best illustrated by the opera, which dates to the late sixteenth century. The visual aspects of musical performances in general include the presence, movements, and gestures of musicians and vocalists as well as uniforms, shiny instruments, marching feet, formal attire, elaborate costumes, wild gyrations, etc. In some performances the exertions and energy required of performers to put on the visual stage show are so extreme that they could not simultaneously be producing the music going out over the loudspeakers.

Music, ranging from subtle background pieces to featured vocal numbers, has been a ubiquitous element of movies since the invention of "talkies." Even in the silent-film era, screenings included musical accompaniment as an integral part of the show. Although **music videos** have come to include just about any imaginable combination of sound and sight, early such short segments that later were recognized as being music videos tended to feature popular music with action shots of performers. A quintessential example is "Can't Buy Me Love," a song performed by the Beatles as part of their 1964 movie, *A Hard Day's Night*, and released that same year as a single and as part of an album of songs from the film. Some twenty years later it was repackaged as a stand-alone music video. At the first MTV Video Music Awards program in 1999, the director of the movie was credited with having "invented" the music video. This attribution, like that accorded many other inventors, is an oversimplification (cf. the above discussion of the invention of the telephone and that of the steam engine). During the 1970s a number of TV programs that included music videos were being aired, and proposals for channels devoted entirely to such content were put forward. But lasting success for such channels was not achieved until 1981 when MTV was launched via cable and satellite.

Another important innovation in communication methods is the use of **optical-fiber cables**. These are made up of extremely thin, long, and flexible fibers of glass, silica, or plastic, through which light signals are sent. Because the frequencies of electromagnetic waves in the visible-light range are so much higher than those utilized in traditional radio and television communication, the amount of information that a single optical fiber can convey is many thousands of times greater than can be carried by a copper-wire circuit. It is crucial to this new process that the light signals travel along inside the fibers—even when they twist and turn—rather than passing out into the space that surrounds them. In general terms, light travels in straight lines without changing direction unless reflected or refracted at a boundary between mediums (i.e., between air and water, air and glass, water and glass, etc.). When light is incident upon such an interface, that fraction of it that passes into the second medium will travel (to varying degrees) in an altered direction (a process called "**refraction**"). This fact is exploited in making lenses for spectacles, telescopes, cameras, etc. What is not generally recognized is that some of the light incident upon such a boundary will be reflected back into the first medium. Indeed, under certain circumstances, all of it will be. Such an effect is referred to as "**total internal reflection**." Learning to exploit this phenomenon fully was the key to the development of a powerful new method of sending information from one place to another.

It has been known since the middle of the nineteenth century that light projected up through columns of water (as in fountains) or lengthwise through long glass cylinders does not disperse from their lateral surfaces into the surroundings but propagates (exclusively) up through the columns or within the cylinders until it fades out. This phenomenon is the direct result of two basic laws of optics: (i) the **law of reflection** (according to which the angle of reflection is equal [in magnitude] to the angle of incidence) and (ii) **Snell's law of refraction** (which sets out how the direction in which light is propagating changes when it crosses the boundary between one medium and another). It follows from the law of reflection that if light traveling inside a glass cylinder *roughly* parallel to its lateral surface is incident upon and reflected by that surface, its direction of propagation will continue to be *roughly* parallel to that lateral surface. Just as is the case with the bounce of a table tennis ball (ignoring the effects of spin), the only change in the trajectory of the light when it is reflected is the reversal of any movement there may have been (directly) toward the reflecting surface. There will be no change in its motion parallel to that surface.

The case of Snell's law of refraction is a little more subtle. Its quantitative form is

$$n' \text{ sine } (r) = n \text{ sine } (i)$$

where i is the angle of incidence and r is the angle of refraction (both measured with respect to the perpendicular of the surface marking the boundary between the two mediums) and where n is the **index of refraction** of the first medium and n' is that of the second medium. Basically, the index of refraction of a substance is a measure of how much the direction of propagation of light changes when it crosses the boundary between the substance and another medium. As a reference point, the index of refraction of a vacuum is assigned a value of 1. That of air differs from it by less than one part in a thousand. The index of refraction of water is 1.33. Those of glass or plastic lie in the range from a little less than 1.5 to a bit more than 1.6.

It follows from Snell's law that a ray of light exiting a horizontal slab of glass will have its direction of propagation shifted away from the vertical; i.e., the angle of refraction of the ray will be greater than its angle of incidence (both angles being measured with respect to the vertical). Substituting (into Snell's equation) 1.5 for the index of refraction of the first medium and 1 for that of the second medium (air) yields

$$\text{sine } (r) = 1.5 \text{ sine } (i)$$

In general terms, this result means that increases in the angle of incidence will be accompanied by greater increases in the angle of refraction. The crucial point for present purposes is that for angles of incidence greater than 42° no light at all will emerge from the slab of glass (total internal reflection) because the corresponding angle of refraction would be more than 90° (from the vertical), i.e., in a direction back into the glass slab. Of course, Snell's law applies only to light that does emerge into the second medium. Light arriving at the surface of the glass slab at an angle of incidence of more than 42° simply will be reflected back into it in accordance with the law of reflection. Similar considerations apply to light arriving (from within) at the lateral surface of a glass cylinder. Of special relevance to the present discussion is the case of light that is propagating within a glass cylinder in a direction roughly parallel to its lateral surface. Since its angle of incidence (measured with respect to a perpendicular to the lateral surface) is close to 90°, all of this light that is incident on the lateral surface of the cylinder will remain within it.

Initially the use of glass to convey light signals was extremely limited because the range of such transmissions was so short—no more than a few yards. This limitation did not prevent the early development of devices of this kind for examining the interior of the human body. But it did hold up the use of light as a means of communicating images and information over significant distances. Surmounting this obstacle took time and the efforts of an entire array of scientists and inventors. The main reason for the short range of these transmissions was the presence of impurities in even the highest quality glass available at the time. These foreign elements absorb or otherwise disrupt light signals being conveyed by the glass. By the 1960s concerted efforts were being made at several research centers to find ways to put glass to work in the communications industry, particularly to develop techniques for making glass that is orders of magnitude freer of impurities than any then available. In 1970 the Corning Glass company announced a breakthrough: glass of the requisite purity.

Another means of enhancing the transmission of light signals is "**cladding**," encasing each glass filament or fiber with a transparent material having an index of refraction that is less than that of glass. This arrangement satisfies the conditions for total internal reflection of light traveling along inside a filament or fiber. A third ingredient in the fiber-optics success story is the bundling together of large numbers of optical fibers into cables. These fibers are so tiny that it is possible for optical cables to encompass hundreds or even thousands of them, providing an enormous capacity to communicate information. Yet another key element of optical-fiber technology is the **semiconductor laser**. Not only do these devices provide an ideal source of light for carrying information via glass fibers, they are very energy efficient and can be operated at room temperature without any special provision for cooling. Lastly, the Herculean communication capacities of optical-fiber cables could not have been put to general use without data-processing equipment, at both the sending and the receiving end, made possible by semiconductor technology. The resulting reduction in the cost of such equipment removed the last barrier to the creation of a communication network now so taken for granted that few appreciate what an achievement it was. Given the advantages of fiber optics (enormous capacity, lower costs, energy efficiency, durability, safety, etc.), it is no surprise that this new technology has rapidly replaced older ones for telecommunication purposes.

A third major new method of transmitting information introduced in the latter part of the twentieth century uses **cellular networks** of microwave towers. The significance of this new system is *not* that it first made possible two-way communication by means of electromagnetic signals when one or both of the participants is in transit. That was being done (e.g., from one ship to another) as early as 1920. What cellular networks did was to make it possible for extremely large numbers of people to enjoy this same capability—without interfering with each other's transmissions or with those of radio and television stations, air-traffic controllers, police and fire departments, etc. Prior to the development of these networks, radio was the only method available for two-way communication with someone in transit and some distance away. But that involves powerful transmitters which are expensive, require extensive sources of energy, and give rise to interference problems. In the 1930s, government agencies were set up to bring order to the radio-transmission sphere. These offices set aside bands of frequencies for different purposes (such as law enforcement, national defense, air-traffic control, commercial broadcasting, etc.) and assign specific frequencies and power levels to individual radio and television stations. Fitting mobile radio-telephone service for the public into this scheme proved difficult.

Once again, as is so often the case with major technological innovations, the currently ubiquitous, hand-held electronic communication device emerged over a period of time in a series of halting steps that range from technical feasibility studies to functioning systems providing service limited to a certain group or location, such as a police department, an urban area, or a transportation corridor. One early (1964) innovation, which might be said to mark the transition from radio-telephone to mobile-telephone service, was the introduction of equipment that provides for simultaneous two-way conversations and eliminates the need for the push-button used (in radio-telephoning) to switch from the transmit to the receive mode or vice versa (changes signaled by the phrases "over," "over now," and "over and out"). The crucial step, however, was the successful implementation of the cellular approach, which features a network of low-powered transmitters, the signals of which reach only very small areas called "cells." Only when a mobile telephone is within one of these small areas does it receive signals from the transmitter there. And only then do antennas at that site pick up the transmissions from a mobile phone. When a user crosses the boundary between one of these cells and another, his or her call is "handed off" to the latter. Elaborate circuitry and complex switching systems are required to "keep up" with which cell a user is in and to shift his or her call (both the incoming and the outgoing parts) to the next cell when the time comes. Initially these systems were operated manually in essentially the same way that long-distance telephone calls were once handled by operators. The result, of course, was that the service was very expensive and its use restricted; but that was soon to change.

Advances made during the 1970s in two areas were crucial to the success that cellular communication ultimately was to have. Computers became much more powerful, affordable, and widely available. Secondly, major improvements were made in the quality of optical fiber (as discussed above). It would be hard to imagine a task more suited to computerization than changing the connections of two circuits when a signal arrives indicating that they, out of thousands of such pairs, need to be changed in a certain way. Given the advantages that computers bring to cellular communication systems, it can hardly be surprising that by the late 1970s a fully functional cellular-telephone system incorporating dedicated computers had been set up or that by the mid-1980s cellular-telephone systems were being set up essentially everywhere. Since rechargeable batteries like those taken for granted by current users of hand-held devices were not available initially, "car-phones" (as they were called) depended on the electrical system of the user's vehicle for energy. Optical-fiber cables are another key element of the elaborate system that now gives an entire population (near-instantaneous) access to an incredible variety of services (available via portable devices). Although it may come as a surprise to many, if not most, users of these devices, all but a small fraction of the path by which the content they receive or send via the cellular system is solid material—not air or empty space. Almost since the beginning, optical-fiber cables have been used to link these cells to central switching facilities and, in turn, to link those facilities in a globe-spanning network. Because of their remarkable qualities, reviewed above, cables of this kind are ideal for the purpose.

The dawning of the twenty-first century witnessed the appearance of yet another marvel, the "**Wi-Fi hot-spot**," a relatively small area within which enormous amounts of data can be sent or received by mobile devices extremely rapidly as well as wirelessly. As is the case with the cells of mobile-phone systems, a central feature of hot-spots is a transmitter/receiver (called a "base station" or "**wireless access point**"), the signals from which reach only a limited area (setting the boundaries of the hot-spots, just as happens with cells). In both cases, these devices are connected to the worldwide telecommunications network by optical-fiber cables that can deliver or receive and forward vast quantities of data at a prodigious rate. The challenge is to move (wirelessly and exceedingly rapidly) these enormous amounts of data between the wireless access points and the mobile devices of a multitude of clients while using only a limited part of the electromagnetic spectrum and avoiding these operations interfering with each other. Part of the solution is to limit the range of the electromagnetic transmissions involved even more than is done in the case of cell-phones—indeed, much more. Severely limiting this range—and thus the size of the hotspots surrounding each wireless access point—while reducing the distance between the access points and increasing their overall number adds to the capacity of the system. This same strategy was used by cell-phone companies in areas with dense populations and when smart-phones were introduced. Despite wide variations in both cases, the two systems differ in that hot-spots typically are some two orders of magnitude smaller than the cells of mobile-phone networks. And just as mobile-phone service can be offered in areas much larger than a single cell (by setting a grid of microwave towers), Wi-Fi can be made available in areas larger than that provided by a single wireless access point by setting up a number of them adjacent to each other.

H. Impact on the Music Industry, Listeners, and Producers

The three technologies discussed above (satellites, optical-fiber cables, and cellular networks), together with incredibly powerful miniaturized computers, provide the foundation for the ongoing transformation that so many areas have been undergoing for the last quarter of a century. Especially noteworthy is the impact of these developments on popular music: the way it reaches the ears of listeners, the way the sound-wave patterns are produced, and the way new musical compositions are created. And there have been major consequences for the music industry, performers, composers, and lyricists.

The new order is epitomized by the capabilities of the portable electronic devices that are now so taken for granted. Particularly important for purposes of this study is the ability to access, essentially at will, a vast smorgasbord of musical recordings either by (i) downloading files onto a portable device, (ii) streaming (listening to music as the signals arrive from a selected source), or (iii) availing oneself of collections of recordings that one has assembled in the cloud. Significant steps along the way include the digitalization of cellular-phone systems (1991 onwards) and the invention of the **smartphone** (1993). As noted above, there was some posting of digitized musical recordings on the World Wide Web following the development of MP3 in the mid-1990s. But its impact on music was limited prior to the turn of the century. Early on, few people even knew that MP3 existed, and only some had access to it. A further impediment to the unrestrained downloading of files of digitized music was the relatively meager data-storage capabilities of the early smartphones, which limited them to about twelve songs. The introduction of magnetic hard drives for portable devices in late 1999 boosted this capacity by two orders of magnitude. Within two years the Apple iPod, a very popular multi-purpose pocket-computer widely used to download digital music files as well as to listen to music, made its appearance.

An immediate effect of the new technology has been to change fundamentally the economics of the music industry. The threat it posed to the existing order became evident following the development by two college students of a software package (using MP3 as a file format) that enabled music aficionados to share their digital files. Just as Berners-Lee (b. 1955) made it possible for participating scientists to access information stored on each other's computers, so these students made it possible (beginning in 1999) for participating music-lovers to share their digital recordings. "**Napster,**" as the service was called, was wildly successful; at its peak there were some 26 million subscribers. In defense of their financial interests, performers and purveyors of musical wares went to court, accusing the service of contributing to copyright infringement. The result was that Napster was forced to cease operating in July 2001. But this legal victory did not return the music industry to its former state. Fundamental changes in the economics of the music sector were already underway. CD sales were in a steep decline that, with minor interruptions, was to continue into the second decade of the twenty-first century.

Although the unauthorized copying of digital recordings continued at a brisk pace after Napster folded, entrepreneurs in growing numbers seized upon the situation as an opportunity to market digital music on the internet, using a variety of business models. A pioneering such enterprise is Apple's iTunes Music Store, launched in 2003. Having entered into contracts with the producers of recorded music, it offers (for a fee) downloads from a vast collection of recordings: singles, albums, releases of extended length, etc. The foundations of a quite different business model for purveyors of musical wares were laid two years later when YouTube made its debut. The resounding success of this venture demonstrated the viability of platforms that supply content in a continuous stream to users who view or listen to it as it arrives without down-loading or archiving. In that regard, this process (which later came to be called "**streaming**") resembles TV and radio broadcasts in some ways but is otherwise very different. When streaming, the user selects the item to be played and can interrupt it at any time as well as rewind or fast forward it. YouTube was launched to make it possible for individuals or small groups who have created video clips to place them before the eyes of viewers. Over time, users of the service posted a great variety of recorded materials on it, including audio files.

The growing popularity of listening to music by streaming inspired entrepreneurs to introduce services providing music in that format. Again, business models vary. They include free offerings supported by advertising revenue, content provided to paying subscribers, as well as combinations of these. Some subscription services have offered initial free trial periods. These enterprises operate under arrangements with copyright owners, producers, and performers, who are compensated according to the extent to which their products are used. Beginning in 2005, high-tech companies began offering "**cloud services**," providing subscribers with (essentially) unlimited storage space on their servers as well as programs for them to use in managing their files. Because of the economies of scale and the convenience provided, as well as the advantages of using a smaller, less expensive mobile device (or on-site operation), these services have experienced explosive growth. Other business models continue to appear, such as combinations of radio stations (supported by advertising) and apps that allow subscribers to replay songs heard on the radio and to download them for future enjoyment.

This shift among consumers of musical products, from acquiring (physical) records of performances to obtaining access to vast collections of electronic versions, has led to profound changes in the paths traveled by performers to reach stardom. When major recording studios were profiting handsomely from the sale of records, tapes, and CDs, they sought out promising new talents and promoted their songs and albums. But as streaming has become more and more popular, the sale of such wares has declined. While the albums of older, well-established rock stars have continued to sell in large numbers, younger performers, even those who have attained superstar status, find that most of their income derives from touring.

This situation poses a special challenge for less-well-known performers, as they are not likely to fill large venues. Among the options available to them is participation in festivals, which are smaller, have lower-priced tickets, and are more likely to attract younger audiences with limited disposable incomes. Additional alternatives stem from the new technology itself. Some platforms provide opportunities for aspiring performers to get their music before sufficiently large audiences for them to have a chance of becoming known on the national and international scene. And some who have posted their music on the web have been "discovered" by large-scale purveyors of musical wares and gone on to stardom. Others have gone into the business of marketing their own recordings on the web.

Another group that has not been well served by the transition to the new system consists of the song writers, composers, and others similarly involved in the production of music. Their level of compensation remains significantly below what it once was. Contributing further to this unpromising outlook is a largely unanticipated trend among aficionados of popular music. As seen above, the new communications technology has raised the number, variety, and accessibility of musical works to a whole new level. Understandably, this led to the expectation that listeners would take advantage of the opportunity to explore musical genres that were unfamiliar to them, as many undoubtedly have done. And one can reasonably assume that some of them have developed an appreciation for one or another new genre and even to have changed the way they evaluate musical performances. Even so, the actual trend has been for listening to become more and more concentrated on the most popular numbers and the most prominent platforms. Paradoxically, the great increase in choices made possible by the new technology has been followed by a narrowing of the range of choices actually made. While the reasons for this result are surely complex and subtle, one thing is clear: it is, if anything, harder than ever for aspiring young performers to find a place in the spotlight and for other workers in the area to receive adequate compensation.

Even more significant for the latter group are the fundamental changes that have come about in the way that new songs on track to become hits are produced. Rather than being created by a single tunesmith or by a composer and a lyricist working together, nearly half of the songs that now make *Billboard*'s "Hot 100" chart have been fashioned by a team of four or more members. At the heart of this development is what is called the "**track-and-hook**" method of producing popular songs. Computers are essential to this process. They provide storehouses from which material can be drawn, a site at which it can be manipulated and processed, and the means to do so. The first step is to use specialized software to search databanks for suitable musical patterns and, on that basis, to synthesize an instrumental track. Other team members then add **hooks**, those segments of popular songs that have a particularly memorable rhythmic, melodic, and lyrical character. Examples from an earlier time are "Stop in the name of love," "You ain't nothin' but a hound dog," and "I want to hold your hand." As more melodic and lyrical phrases are added to the mix, computer programs are used to adjust their qualities (e.g., pitch and rhythm) and to move them about in the lineup ("cut-and-paste"). Lastly, in order to gauge its potential for success, computers are used to compare the sonic patterns of the new song to those of earlier ones that became hits.

In contrast to developments set in motion by the new technology that tended to make it harder for aspiring performers to attract attention, such platforms as YouTube have made it possible for individuals, small groups, and organizations as well as businesses to upload content of just about every imaginable kind to the World Wide Web with the potential to "go viral." Most notable for present purposes, this development opened the door to the return of *Amateur Hour*, albeit in a new format: the music video. The digitization of sound and visual images made it possible for them to be manipulated and combined in an almost unlimited number of ways. Software that has become available enables almost anyone to create videos and to upload them to the web. So popular has viewing these creations become that those who excel at crafting them have attracted huge followings. Consequently, video-sharing websites such as YouTube are willing to pass on some of their advertising revenue to those who post popular videos. While business models vary, these payments are usually based on such factors as the number of hits a video attracts or the length of time it is viewed in combination with advertising.

* * *

The above accounts demonstrate in concrete terms the profound effects that technological innovations can have on the development of an area of human activity and/or interest. These effects include making a product or service more available to more people, more of the time, under more kinds of circumstances, and in a greater variety of forms (certain ones of which may involve fundamental changes in the product or service). Inevitably some such innovations are disruptive and are likely to have a greater impact on this or that group than on others. The availability of powerful new technologies can affect the choices people make, their preferences, and the standards they employ in making aesthetic and ethical assessments. But despite the advantages of a new technology, people do not necessarily completely abandon older devices and techniques. While this is true in other areas, it is especially significant in music. People continue to play and sing before audiences of varying sizes with minimal or no reliance on recent technologies. Despite the success of the invention of the electric guitar, acoustical guitars are still in use, and the sale of LPs is increasing (to aficionados dissatisfied with digital recordings, especially MP3 versions). Although the impact of digital technology continues to be significant in shaping the music industry, the future of music and music technology is in no way decided.

Sources / Further Reading

Alwayn, Vivek. *Optical Network Design and Implementation*. Cisco Press, 2004.
Bacon, Tony. *Six Decades of the Fender Telecaster: The Story of the World's First Solidbody Electric Guitar*. Backseat Books, 2005.
Berners-Lee, Tim. *Weaving the Web: The Original Design and Ultimate Destiny of the World Wide Web*. HarperCollins, 2000.
Braun, Ernest, and Stuart MacDonald. *Revolution in Miniature: The History and Impact of Semiconductor Electronics Re-explored in an Updated and Revised Edition*, 2nd ed. Cambridge University Press, 1982.
Carroll, Michael Thomas. *Popular Modernity in America: Experience, Technology, Mythohistory*. State University of New York Press, 2000.
Craughwell, Thomas J. *The Book of Invention*. Black Dog & Loventhal Publishers, Inc., 2008.
Devine, Kyle. "A Mysterious Music in the Air: Cultural Origins of the Loudspeaker." *Popular Music History*, vol. 8, no.1, February 2014, pp. 5–28. doi.10.1558/pomh.v8i1.5.
——. "Imperfect Sound Forever: Loudness Wars, Listening Formations and the History of Sound Reproduction." *Popular Music*, vol. 32, no. 2, May 2013, pp. 159–76, doi.org/10.1017/S0261143013000032.
Doster, Alexis, editor. *The Smithsonian Book of Invention*. Smithsonian Exposition Books, 1978.
Dyson, James. *A History of Great Inventions*. Edited by Robert Uhlig, Carroll & Graf Publishers, 2001.
"Ed Sheeran: Reinventing Pop Music." *Sunday Morning*. CBS. 5 March 2017.
Einaarsson, Göran. *Principles of Lightwave Communications*. John Wiley & Sons, 1996.
Evans, Tom. *Guitars: Music History, Construction and Players from the Renaissance to Rock*. Paddington Press, Ltd., 1977.
Farmer, Ken. "Appraisal, Bacon Sultana Banjo 1936." *Antiques Road Show, St. Louis*. PBS. 20 June 2015.
Fusilli, Jim. "Remembrances: Les Paul 1915–2009: The Father of Modern Guitar Spanned and Shaped Genres." *The Wall Street Journal*, 14 August 2009, p. A9.
Gaarofalo, Reebee. "From Music Publishing to MP3: Music and Industry in the Twentieth Century." *American Music*, vol. 17. no. 3 Autumn 1999, pp. 318–54.
Giancoli, Douglas C. *Physics: Principles and Applications*, 7th ed. Pearson, 2013.
Greenstein, Shane. "Silicon Valley's Lawmaker: Review of *Moore's Law* by Arnold Thackray et al." *The Wall Street Journal*, 26 May 2015, p. A11.
Hajdu, David. *Love for Sale: Pop Music in America*. Farrar, Strauss and Giroux, 2016.
Halliday, David et al. *Fundamentals of Physics*, 10th ed. Wiley, 2013.

Huber, John. *The Development of the Modern Guitar.* Bold Strummer, 1991.

Jackson, John David. *Classical Electrodynamics*, 3rd ed. John Wiley and Sons, 1998.

Karp, Hannah. "Competition Grows in On-Demand Music: Radio Broadcaster iHeart to Enter Packed Market for Subscription Services." *The Wall Street Journal*, 24 September 2016, p. B4.

——. "Future of Music: Playing for $30,000 in Tips." *The Wall Street Journal*, 27 March 2017, pp. 1–2.

——. "Scores of Music Services Stream into Crowded Field." *The Wall Street Journal*, 24 December 2013, p. B1.

Katz, Mark. *Capturing Sound: How Technology Changed Music*, revised ed. University of California Press, 2010.

Kizza, Joseph Migga, editor. *Ethical and Social Issues in the Information Age*, 6th ed. Springer, 2017.

Leonard, Michael. "Les Paul's Recording Innovations." *Gibson.com.* 23 December 2013, gibson.com/news/-lifestyle/features/en-us/Les-Paul-Recording-innovations

"Lip Syncing." *Sunday Morning.* CBS, 23 February 2017.

Loh, Tim et al. "Rare Earths: Commodities Bust: Scarcity Proved to Be a Mirage." *Seattle Times*, 3 July 2015, pp. A8+.

"Magnetic Tape Recorders and Reproducers—'Recording with tape.'" *HistoryofRecording.com.* www.historyofrecording.com/Magnetic-Tape-Recorders.html

Malone, Michael S. "The Lion of Silicon Valley: Escaping from Communist Hungary and Building an American Business Career That Changed the World." *The Wall Street Journal*, 23 March 2016, p. A11.

——. "The Promise at Technology's Powerful Heart." *The Wall Street Journal*, 18 April 2015, p. A11.

"The Paradox of Choice." *The Economist*, 11 February 2017, p. 12.

Pirtle, Caleb III. *Engineering the World: Stories from the First 75 Years of Texas Instruments.* Southern Methodist University Press, 2005.

Reid, T.R. *The Chip: How Two Americans Invented the Microchip and Launched a Revolution.* Simon and Schuster, 1984.

Robjohns, Hugh. "A Brief History of Microphones." *Microphone-data.com.* www.microphonedata.com/media/filestore/articles/History-10.pdf. First published in *Microphone Data Book*, edited by Chris Wolf. Human-Computer Interface, 2001.

Rosenberg, Paul. "The Basics of Fiber Optics — Part 1. *ECM.com.*, Informa USA, Inc., 1 May 1999. ecmweb.com/basics/basics-fiber-optics-part-1

Stoll, Clifford. *Silicon Snake Oil: Second Thoughts on the Information Highway.* Doubleday, 1995.

Taylor, Timothy D. *Strange Sounds: Music, Technology & Culture*. Routledge, 2001.
—— et al., editors. *Music, Sound, and Technology in America: A Documentary History of Early Phonograph, Cinema, and Radio*. Duke University Press, 2012.
Trust, Gary. "Why Solo Songwriters Are No Longer Today's Hitmakers." *Billboard.com*. www.billboard.com/articles/news/6738318/why-solo-songwriters-are-no-longer-todays-hitmakers. Originally published in *Billboard*, 31 October 2015.
Turner, Richard, and Steve Dougherty. "SiriusXM's Battle for the Car Radio." *The Wall Street Journal*. 19 December 2016, p. A15.
Van den Hover, Jerven, and John Weckert, editors. *Information Technology and Moral Philosophy*. Cambridge University Press, 2008.
Walrand, Jean. *Communication Networks: A Concise Introduction*. Morgan and Claypool Publishers, 2010.
Wheeler, Tom. *The Stratocaster Chronicles*. Hal Leonard, 2004.
White, Forrest. *Fender: The Inside Story*. Backbeat, 1994.
Williams, Trevor. *The History of Invention: From Stone Axes to Silicon Chips*, revised ed. Checkmark Books, 2000.
"Wurlitzer Organs and Other Theater Organs." *Sunday Morning*. CBS. 14 June 2015.

Postscript:
The Digital Age and the Future of Technology

CENTRAL TO THE VERY nature of technology is its close connection to human inventiveness, which manifests itself in the last of the five defining characteristics set out in the second chapter of this book: ongoing innovation. As demonstrated many times throughout this text, inventions often make possible additional inventions. Consequently, the greater the number of inventions that already exist, the greater the number that are likely to be devised: the "snowball effect." Historically, the extent to which this takes place depends upon the conditions prevailing at a given time, most especially upon whether or not people have a permanent or semi-permanent place of residence. Foragers who move about in search of fresh supplies of food have a technological repertoire that is limited to what they can carry with them or practice on the spot. While there have been cases in which non-agricultural peoples have had permanent residences, the practice of agriculture imposes a more or less sedentary lifestyle on its practitioners. Thus, the birth of agriculture marks the beginning of the ongoing, if irregular, process of accelerating technological innovation that continues right up to the present. Major milestones include the harnessing of the muscle power of animals, water power, and steam power, as well as the literacy revolution and the formation of the technology-and-science partnership. Most recently, the frontier lies in the area of computers and communication technology, focusing on such topics as AI (artificial intelligence). Where it will lead is an open question.

As transformative as the technological innovations of the late nineteenth and twentieth centuries were, their impact pales in comparison to that of more recent "high tech" innovations. Few if any areas of human activity or interest surveyed in this book have escaped being affected significantly by computer and information technology. We continue to see the powerful partnership of science and technology in the application of quantum mechanics to enhance computing. The science exploring how proteins fold (which, when folded incorrectly, can give rise to diseases like Parkinson's or Alzheimer's) has received a massive boon from computer simulations that can help predict the many ways proteins can (mis) fold. And the mutual influence of technology and the economy is clearer than ever. Over the past decade, the percentage of tech companies comprising the S&P 500 grew from approximately 20 percent to almost 50 percent. Cryptocurrency has given rise to new economic currencies and speculation entirely undergirded by computer and information technology. The ever-increasing ubiquity of social media impacts our society in countless ways, from spreading (mis)information to allowing immediate social access to others across the globe at an unprecedented scale. Moreover, the data centers that are required for the ongoing function of the internet, apps, phones, and large language models are producing massive carbon emissions and using enormous quantities of water for cooling, impacting the environment and contributing to climate change. Yet computer technology is also indispensable to our ability to monitor and model climate change, helping us understand the danger that the planet is in and the steps we might take to mitigate that damage.

Despite its overarching snowballing pattern, despite the achievements of its partnership with science, and despite its current dizzying pace, technological innovation certainly has not always been underway everywhere and at all times. Its flourishing at some place at one time is no guarantee that it will continue to do so. Predicting the future is an inexact science—if it is a science at all. With a few notable exceptions, prognostications regarding technology are notorious for being wrong.

Sources / Further Reading

Dyson, James. *A History of Great Inventions*. Edited by Robert Uhlig. New York: Carroll & Graf Publishers, 2001.

James, Peter and Nick Thorpe. *Ancient Inventions*. New York: Random House, Inc., 1994.

Mach, Ernst. *The Science of Mechanics: A Critical and Historical Account of Its Development*, 6th ed. Translated by Thomas J. McCormack. La Salle, Illinois: The Open Court Publishing Company, 1960.

Plutarch. *The Lives of the Noble Romans and Grecians*. Translated by John Dryden, edited by Arthur Hugh Clough. Chicago: Encyclopedia Britannica, 1952.

Index

About the Publisher

The word "broadview" expresses a good deal of the philosophy behind our company. Our focus is very much on the humanities and social sciences—especially literature, writing, and philosophy—but within these fields we are open to a broad range of academic approaches and political viewpoints. We strive in particular to produce high-quality, pedagogically useful books for higher education classrooms—anthologies, editions, sourcebooks, surveys of particular academic fields and sub-fields, and also course texts for subjects such as composition, business communication, and critical thinking. We welcome the perspectives of authors from marginalized and underrepresented groups, and we have a strong commitment to the environment. We publish English-language works and translations from many parts of the world, and our books are available world-wide; we also publish a select list of titles with a specifically Canadian emphasis.

broadview press

This book is made of paper from well-managed FSC® - certified forests, recycled materials, and other controlled sources.

FSC www.fsc.org MIX Paper | Supporting responsible forestry FSC® C103567

PCF

BIO GAS ENERGY

PERMANENT